PERCY, CLARENCE, QST AND THE ARRL

Before, During and After the Birth of
Amateur Radio

Bitterroot Mountain
PUBLISHING, LLC.

Hiram Percy Maxim "The Old Man"

PERCY, CLARENCE, QST AND THE ARRL

Before, During and After the Birth of Amateur Radio

As told by
Larry Telles
K6SPP

Bitterroot Mountain
PUBLISHING, LLC.

Percy, Clarence, QST and the ARRL: Before, During and After the Birth of Amateur Radio.
Copyright © 2014 Larry Telles
Published by Bitterroot Mountain - 9030 N. Hess Street, Suite 331, Hayden, ID 83835
Visit our Web site at www.BitterrootMountainPublishing.com
First edition: 2014

Front cover and interior design by Larry Telles.
ARRL Diamond is the official logo of the American Radio Relay League.

Photographs: All images indicate the source. Those without a source are from the author's collection.

This is a work of non-fiction. Although the author and publisher have made every effort to ensure the accuracy and completeness of information contained in this book, we assume no responsibility for errors, inaccuracies, omissions, or any inconsistency herein. Any slights of people, places, or organizations are unintentional.

ISBN 978-1-940025-13-1
Library of Congress Control Number: 2014906512

Printed in the United States of America.

1. 100 Years of ARRL. 2. Hiram Percy Maxim. 3. Clarence D. Tuska. 4. Ham Radio. 5. The Old Man. 6. American Radio Relay League. 7. QST Magazine. 8. Newington, Connecticut. 9. Spark Gap Transmitters. 10. CQ.

This book is dedicated to the next
generation of amateur radio operators,
both young and old. It is their duty to
carry on into the 21st
Century providing enjoyment for
themselves and service to others in
the spirit which has always been that
of amateur radio.

Preface

Most people both young and old who involve themselves in amateur radio usually have different reasons. Once licensed by the FCC, operators begin to gain experience and generate thousands of good stories transcending the years. From the time I was introduced to amateur radio, it took nine years for me to get that FCC license a.k.a. a ham "ticket." It all started with my grandmother who worked as head linen lady at the Shattuck Hotel in Berkeley, California. One day in 1949 she brought home a copy of a QST Magazine that someone had left behind in one of the hotel rooms. I was eleven at the time. I found the magazine amazing. It didn't come closed to Buck Rogers or Flash Gordon, because it was for real and I could actually participate in the science discussed. I read that magazine off and on for years.

By October of 1957 I had just gone to work for the telephone company in Oakland, California. One night while working a swing shift I went to the break room for dinner. There sat one of my co-workers reading a QST Magazine . I don't remember what I had for dinner that evening, but I sure remember our conversation. I assembled all the information necessary to study the Morse code and radio theory and on April 15, 1958, in a very cold room at 555 Battery Street, San Francisco, the FCC District Office, I took and passed my Technician Class Radio Amateur license. I was a "Ham" at last!.

Over the years I had read bits and pieces about the ARRL but never looked much at its history. When I volunteered to give a talk at the 2014 Boise ARRL Hamfest gathering, my research for my *PowerPoint** presentation began. The ARRL organization and its history was a complex but fascinating story about a truly wonderful association of Hams who were able to survive every challenge it encountered, none the least of which was World War My entire research from that presentation to today is embodied in this work.

* Powerpoint is a product on the Microsoft Corporation.

Table of Contents

Introduction

You can't jump into "wireless telegraphy" without first dealing with the regular wired kind. It isn't necessary to go all the way back to Ben Franklin to learn about electricity. However the discovery of the wireless telegraph can be traced back to a kite similar to Franklins in 1811. This material is going to look at what came before the obvious components of the telegraph...the electromagnet. In the world of telegraphy there is still controversy today as to who did what first. There are many human demonstrations, inventions and experiments taking place simultaneously making the electromagnet possible. It was during the time the telegraph was being established that a group of young men started experimenting with spark. With no rules and regulations at first, the battle between citizen and military began. The ARRL with Hiram Percy Maxim's leadership that helped build the foundation of amateur radio.

Chapter 1: Those who came before

The battery is not directly related to the electromagnet, but it shows how early people were interested in electricity in the 1800s. Italian born Alessandro Giuseppe Antonio Anastasio Volta (1745-1827), was its inventor. His invention became known as the voltaic pile which consisted of pairs of zinc and copper stacked on top of each other. Each layer was separated by a layer of cardboard or cloth, soaked in brine as the electrolyte. [1]

Antonio Volta

A very crude telegraph was invented by Samuel Soemmering (1755-1830) in 1809 Bavaria. Thirty-five wires were used with gold electrodes emerged in water. Two-thousand feet at the receiving end the message was read by the amount of gas created by electrolysis. It doesn't sound practical. [2] A few years later in 1811, Soemmering took a different approach. He replaced wires with water and successfully transmitted telegraphically across basins in his laboratory. [3]

S. Soemmering

One of the electrical properties used in the electromagnet telegraph system came from Hans Christian Oersted (1777-1851). He was a Danish physicist who in 1820 discovered that when electric current travels through a wire it generates a magnetic field. This field can deflect a compass needle. [4]

Hans Oersted

The first time that the electromagnet was introduced was 1825. British inventor William Sturgeon (1783-1850) laid out the foundation for future developments in electromagnets. His crude device used bare wire, no insulation and the iron core was varnished and was bent into a U-shape. [5]

William Sturgeon

Sometime before 1827 a French physicist and mathematician Hans Marie Ampere (1775-1836) established a statement of electric current. It was the definition of the unit of measurement of current flow, the ampere. [6] Shortly after Ampere made his definition, GeorgSimon Ohm (1789-1854) discovered the mathematical law forelectric-current. The current flow through a conductor is directly proportional to the potential difference (voltage) and inversely proportional to the resistance. He named this Ohms law after himself. [7

Hans Ampere   Georg Ohm

Joseph Henry (1797-1878) took William Sturgeon's electromagnet beyond it crude design and made the device more operational and efficient. Henry used wire that he insulated by winding silk tightly coiled on the iron core. He didn't patent the device in 1828, because he believed in the dignity of science and that he shouldn't profit from it. This might be one of the reasons in 1893 why Joseph Henry's name was given to the standard electrical unit of inductive resistance, the "Henry". [8]

Joseph Henry

Harrison Gray Dyar (1805–1875) an American chemist and inventor sent electrical sparks through chemically treated paper tape to burn dots and dashes. The paper was a ribbon of moistened litmus paper on a spool that revolved mechanically by a clockwork apparatus. Nitric acid formed on the litmus paper by the action of the electricity left appropriate and legibly small red marks for designated letters. Dyar's method was of frictional electrolytic nature where Morse's was an electromechanical usage. This was performed in late1828. Dyar did erect the first telegraph line, transmitting over it the first telegraph message ever sent in America, but no electromagnets were used in his operation. [9]

What was bound to happen did. Three individuals came forward with a working telegraph system. In 1830 Joseph Henry demonstrated William Sturgeon's device that he had improved. Henry sent an electronic current to the far end which activated an electromagnet causing a bell to ring. The distance involved was over a mile. Two British physicists, William Cooke and Charles Wheatstone were the second team to come forward with a telegraph system employing an electromagnet. However, it was Samuel Morse (1791-1872) in 1832 who came up with the most practical and commercial operation. [10]

In the spring of 1837 Morse began to improve his telegraph by sending a message through ten miles of wire. He takes on a partner for financial assistance. Shortly after the partnership began, the partner claims that he invented the telegraph. Morse takes him to court and wins. This would be the first of many court appearances. On September 28, Morse files for a patent for his telegraph. January 1838, Morse makes a big change. He created a dot-and-dash code that used different combinations to represent letters of the English alphabet and the ten numbers. This coding system was far superior since it didn't require coding or decoding, but could be "sound read" by operators. Later in 1838, at an exhibition of his telegraph, Morse transmitted ten words per minute using the new Morse code that would become standard throughout the world. It wasn't

Samuel Morse

until 1840 that Morse received his patent. October 1842 Morse tried underwater transmission between two islands in New York harbor, a distance of two miles. The demonstration was a big success. Acceptance of the Morse telegraph began on March 3, 1843 when Congress appropriated $30,000 for an experimental telegraph from Baltimore to Washington D.C. The cable that carried the signals was installed and placed in underground lead pipes several months earlier. May 24, 1844 Morse sent a message from the Capitol building in Washington, D.C., to the Railroad Depot in Baltimore. With the the now famous message, "What Hath God Wrought?" The telegraph line was expanded in 1846 from Baltimore to Philadelphia. While Washington D.C., is connected to New York. New telegraph companies began to materialize and Morse's patent claim was threatened once again. In 1854 the U.S. Supreme Court upholds Morse's claims for his telegraph system. "He must be paid when companies use his system." [11]

With the common telegraph system up and running, inventors turned to wireless operation which will soon be called radio. [12] It appears that over the past centuries there have been several definitions of the meaning of "wireless telephony." If you were an observer in the sixteenth century you may have thought "sympathy" (an invisible force) existed between needles touched by the same magnet, because a deflection of one would cause a corresponding deflection of the other. This thought process came before, during and after the electromagnet arrived on the scene. [13]

In 1854 James Bowman Lindsay (1799-1862) received a patent for his wireless telegraphy system that sent signals through water. The work he demonstrated never got out of the experimental stage. In 1860 Lindsay successfully sent a signal across the River Tay in Scotland. He died two years later still believing he had a sound idea. [14]

When attention was focused on wireless, a large merger took place. The New York and Mississippi Printing Telegraph Company joined with other smaller telegraph companies to form the Western Union Telegraph Company. [15]

James Linsay

James Clerk Maxwell (1831-1879), a Scottish physicist, predicted the existence of radio waves in the early 1860s. This together with Heinrich Rudolph Hertz (1857-1894), a German physicist in 1886, established the theory that rapid variations of electric current could be propagated into space in the form of radio waves similar to those of light and heat. [16]

Mahlon Loomis (1826-1886), a dentist who had invented and patented a process of making artificial teeth became interested in wireless telegraph. October of 1866 Loomis claimed that he transmitted signals between two Blue Ridge Mountain-tops fourteen miles apart in Virginia, using kites as antennas. However, he could not identify the names of any of the

James Maxwell

witnesses. His system used atmospheric electricity for telegraph communication. Loomis filed his patent (#129,971) on July 30, 1872. With no financing at hand Loomis turned to Congress. A bill was introduced in 1869 and with no vote taken. He got a backer, who later lost all his money on Black Friday, (September 24, 1869). Again in July, 1870, another bill was again introduced in Congress with no action taken. In 1871 Loomis found a couple of backers in Chicago who were willing to put up $20,000 to underwrite the Loomis Aerial Telegraph Company. That all changed on October 8, 1871 when the great Chicago fire burned out the backers. Congress finally decided to vote on the Loomis bill. May, 1872 bill was defeated. The same bill came up on the daily calendar and was passed. Now Dr. Mahlon Loomis had a Congressional charter in one hand and a patent in the other. The year 1873 was a very bad year for business and few had any money to invest. [17]

Mahlon Loomis

April 2, 1872 Samuel Morse died in New York City at age eighty-one. [16] A year later there were estimated to be 150,000 miles of telegraph lines in the United States. These lines were run by Western Union who had successfully bought up most of the smaller companies. [18]

While most telegraphy inventors were still working on wireless, Alexander Graham Bell (1847-1922) was looking to find a way to send multiple telegraph messages on each telegraph line to avoid the great cost of constructing new lines. Message traffic was rapidly expanding. So Bell was trying to create a harmonic telegraph or acoustic telegraphy. One of the multiple set of reeds was accidentally plucked by Bell's assistant Thomas Watson on June 2, 1875. Bell heard the harmonic of the reed at the distant end of the wire. Bell at that moment knew he had developed an acoustic telegraph. He realized that to transmit voice over telegraph wires he would remove the multiple reeds and replace them with one reed or an armature. Bell applied for a patent the morning of February 14, 1876 (patent #174465). Another inventor, Elisha Gray arrived with a similar patent to file that same afternoon. [19]

Alexander
Graham Bell

In 1878, David E. Hughes (1831-1900) discovered that sparks would generate a radio signal that could be detected by listening to a telephone receiver. He developed his spark-gap transmitter and receiver by using trial and error experimentation, until he could send and receive Morse code signals out to a range of 500 meters. One of the witnesses was Sir William Henry Preece. [20]

Nikola Tesla (1856-1943) began working for a telephone company in Budapest, Hungry. That is where he claimed in 1881 to have developed a telephone repeater or amplifier. It was never patented nor publicly described. [21]

Nikola Tesla

Physicist Heinrich Hertz set out to scientifically verify Maxwell's predictions of the existence of radio waves. In 1888 Hertz used a tuned spark gap transmitter and a tuned spark gap detector located a few meters away from each other. In a sequence of Ultra High Frequency experiments, Hertz established that electromagnetic waves were being formed by the transmitter. When the transmitter sparked, small sparks also appeared across the receiver's spark gap. [22]

Sir William Henry Preece (1834-1913) successfully transmitted and received Morse code signals over water for a distance of one mile in 1889. He and Arthur West Heaviside

Heinrich
Hertz

6

conducted experiments in parallel telegraph lines and discovered radio induction. In 1897 Preece began working with Guglielmo Marconi (1874-1937) who sent and received his first radio signal in Italy in 1895. Four years later Marconi flashed the first wireless signal across the English Channel. Then in December, 1901, he received the letter "S", telegraphed from England to Newfoundland. This transmission has been contested on theoretical work as well as a reenactment of the experiment. It is now known that long-distance transmission at a wavelength of 366 meters is not possible during the daytime. This is because the sky wave is heavily absorbed by the ionosphere. It is possible that what was heard was only random atmospheric noise, that he thought was a signal, or could have been a harmonic of the signal. [22] So, the debate goes on.

Sir
William Preece

Chapter 2: Spark-gap and the Hams

The first ten years after the turn of the 20th Century were filled with a multitude of advances in telegraphy and related fields. Some wished to dwell in the past. One such inventor was Nathan B. Stubblefield. He demonstrated in January 1, 1902 his "wireless telephone." Stubblefield claimed he demonstrated the same device, similar to the Edison system in Murray, Kentucky in 1892. It was to be used for moving trains or ship to shore. Few believed that Stubblefield invented radio, except the 1944 Kentucky Legislature who issued a resolution. Stubblefield died in 1928 and never pursued any commercial applications of his invention. [1]

Nathan B. Stubblefield

Theodore Roosevelt

From 1900 to 1904, thousands of radio experimenters followed Marconi's historic event. Around this time about 115 magazines on the subject of wireless telegraphy became available. A protocol for shipboard and coastal stations was signed August 13, 1903 in Berlin. Those signing were Germany, Austria, Spain, the United States, France, Great Britain, Hungary, Italy and Russia. This protocol made no mention of radio amateurs. It was President Theodore Roosevelt on June 24, 1904 who created a board to take into account the increasing problems in wireless telegraphy. Interference was the main problem with competitive jealously between wireless companies being a secondary problem. [2]

The most significant contribution to radio came in 1905. Lee DeForest invented the Audion or three-element vacuum tube. DeForest had always been a dedicated supporter of amateurs. [3] A year earlier, Fleming introduced two-element vacuum valve, or diode called the "Fleming Valve." The term "valve" was used since electrons only flowed in one direction. Fleming's patent was very basic which would be a issue later in a court of

8

law. The development would be in question. Fleming's Valve was announced on October 20, 1906. Seven years after being announced the "Valve now Audion" was not used much. Most historians felt that it came down to cost.

On November 3, 1906 twenty-seven nations signed the International Telegraphy Convention in Berlin. Again amateur radio operators were not mentioned in the signed document. There appeared to be no place for amateur radio. At this body the convention adopted the term "Radio." [4]

John Fleming

Ernst Alexanderson invented an alternator that was designed for long wave radio communications by shore stations. This alternator was too large and too heavy to be installed on most ships. The first of three 50 kilowatt alternators were delivered to Reginald Fessenden at Brant Rock, Massachusetts. On December 24, 1906, together with a rotary spark-gap transmitter and Alexanderson alternator, Fessenden made the first radio audio broadcast, from Brant Rock. Many ships at sea heard his broadcast. Fessenden played *O Holy Night* on his violin followed by reading a passage from the Bible. The other two alternators went to John Hays Hammond, Jr. in Gloucester, Massachusetts and the *American Marconi Company* in New Brunswick, New Jersey. [5]

Ernest Alexanderson

Reginald Fessenden

In the background many amateurs were experimenting with new types of detectors. The crystal detector was the most popular since it added greater sensitivity. Most amateur equipment by 1906 was home brewed. During this time the first successful CW transmitter was invented by a Danish engineer Valdemar Poulsen. His arc transmitter was an earlier idea of English engineer William Duddell who discovered how to make a resonant circuit using a carbon-arc lamp. Since Duddell's "musical arc" operated at audio frequencies. Poulsen found a way to make the arc oscillate at radio frequencies by modifying the electrodes. He placed the arc in an atmosphere of hydrocarbon vapor or

Valdemar Poulsen

9

pure hydrogen, and added a transverse magnetic field. Poulsen's transmitters were used world wide during the second and third decades of the 20th Century. They were replaced by transmitters that were running a vacuum tube as a generator of continuous waves.

Amateur popularity continued to grow. The 1907 magazine, *Electrician and Mechanic* contained a monthly column written by amateurs. The column in the July/August issue was entitled "How to Construct a Simple Amateur Station." Other technical magazines followed. [6]

William Duddell

Price, 10 Cents   September, 1910

ELECTRICIAN & MECHANIC

PUBLISHED MONTHLY BY
SAMPSON PUBLISHING CO.
BOSTON, MASS.

One aspect of the popularity came from a man who's name was Hugo Gernback. Born in Luxembourg in 1884 who came to America twenty years later. During his childhood he developed an interest in electricity. In his early years Gernback coined the word "television," and worked to get amateurs a wide section of the radio spectrum. He started all of this in 1906 by opening a store, *Electro-Importing Company* in New York that sold wireless equipment to the general public. [7] Gernback began publishing a magazine, *Modern Electrics* in 1908. The magazine started out with a general format of electricity and soon shifted mainly to radio. [8]

Hugo Gernback

PRICE 10 CENTS   APRIL, 1908

Vol. I.   No. 1.

MODERN ELECTRICS

Published by MODERN ELECTRICS PUBLICATION, 87 Warren Street, New York, N. Y.

IN THIS NUMBER.

WIRELESS TELEGRAPHY
By Wm. Maver, Jr.
EXPERIMENTS IN STATIC
ELECTRICITY—By J. H. Hooton.
HOW TO MAKE A "DRY" STORAGE
BATTERY FROM A "WET" ONE
By H. Gernback

THE SPEAKING GLOVES.

RECHARGING DRY CELLS.

HOW TO MAKE AN ELECTRIC
WHISTLE.

HOW TO MAKE A MERCURY
INTERRUPTER.

WIRELESS DEPARTMENT.
THE ORACLE.

ELECTRICAL PATENTS
OF THE MONTH.

TECHNICAL NOTES.

"The Electrical Magazine for Everybody"

11

During the year 1908, receiving tuners were the rage with amateurs. Marconi held many patents on both transmitting and receiving filters [tuners]. The tuners used by amateurs were far superior to those used by the U.S. Navy.

With all of the amateurs reading Gernback's magazine he took the next step by forming the first radio amateur radio organization. On January 2, 1909 the Wireless Association of America in New York City was chartered. After only a few months the membership reached 3,200. In November, 1910, according to Gernback, membership had risen to 10,000. This number was fairly accurate since most members were kids, the association officers were adults, with no dues or obligations. To back up his claim, there was an increase in the number of stations on the air about 150+ stations in 1905 600+ by 1910.

Gernback didn't stop with one magazine or organization. In 1909 he published the first *Wireless Blue Book* through his wireless organization. It listed 90 amateur stations that belonged to the association along with call sign, wavelength in meters and spark length. His second Blue Book arrived on June 1, 1910. About this time *Modern Electrics* circulation increased from 2,000 to 30,000 and expanding to 52,000 by March 1911.

During the time of the radio amateur growth, the spark transmitter was gradually improving. Some more well to do amateurs were constructing rigs using high-voltage transformers capable of nearly five-kilowatts. However, the advent of the rotary spark gap was on the horizon.

Advances in equipment didn't make the amateurs problems any easier. Interference was the source of agitation that caused the United States Navy to pressure Congress to act. In 1910, Congress had six bills pending that would clean up the airwaves. One of the six bills would eliminate amateur radio entirely. Gernback used *Modern Electrics Magazine* to express his opposition and in so doing rallied the amateurs. Letters and telegrams of protest flooded Washington which resulting in all six bills being defeated.

Gernback realized that the battle of the amateurs was not over and that government involvement was close at hand. In February, 1912 he laid out his recommendation for the future of amateur radio. First Gernback was of the opinion that amateurs should be confined to 200 meters. That part of the spectrum had long been considered as worthless. He also considered that an amateur's input power should not exceed one kilowatt.

The use of periodicals to gain support for the amateur's cause was gaining in large numbers of supporters. Near the end of the first decade of the 20th Century amateurs vs. the Navy was rising to a fevered pitch. The Navy believed that the only way to get rid of the amateurs once and for all was an Act of Congress. The Burke Wireless Bill was introduced on March 8, 1910, which followed the Depew Wireless Bill, S.7243, submitted on March 6th. Amateur radio operators were not again mentioned by name, but made it illegal for "outsiders" to interfere with the Navy's operation. The Depew Bill was the first to be blasted by editorials in *Modern Electrics Magazine*. The Bill only passed in one house of Congress. Next came a free-for-all. When the dust settled, thirteen bills were defeated in Congress during the first quarter of 1912.

The momentum was shifted for the first time in July, 1912, when the United States sent delegates to the London Conference to ratify the Berlin Convention of 1906. U.S. delegates returned home armed with new detailed regulations concerning the governing of a newly-arrived industry called "radio." Congress took the once defeated Alexander Bill and modified it in accord with the London treaty. Regulation Fifteen specified that amateur radio operators could not use any wavelengths above two hundred meters, except by special permission. The Bill passed Congress on August 9, 1912.

Regulation Fifteen reads as follows:

> No private or commercial station not engaged in the transaction of bona fide commercial business by radio communication or experimentation in connection with the development and manufacture of radio apparatus for commercial purposes shall use a transmitting wavelength exceeding two hundred meters, or a transformer input exceeding one kilowatt; except by special authority of the Secretary of Commerce contained in the license of the station. . . [9]

In order to overcome a language problem between different countries, the Q Code was designed. This affected both ships and shore stations. In 1912 the original list of 50 signals was adopted by international agreement in London. Many of those codes are still familiar to amateur operators today -QRN, QRM, QSO, etc [10]

President William Howard Taft signed the radio act in August 17, 1912, containing the recommendations set forth by Gernback who took full credit and himself the sole spokesman of amateur radio. It was later discovered that Gernback was not alone. Many radio clubs had traveled to Washington to testify in the hearing opposed to the Depew Wireless Bill. That same year the Radio Act was signed at a time when over 122 radio clubs were organized in the U.S. [11]

William Howard Taft

The letters CQ were brought to the English telegraph well over 100 years ago. This signal meant "All stations. The telegraph call CQ was born on the English telegraph over a century ago as a signal meaning "All stations. A statement to all postal telegraph offices to receive the following message." This term originated on the landlines, and was later used by the Marconi Company on radio as a general call to its ships. At the same time other companies were using KA as a general call. The London Convention of 1912 adopted CQ as the international general call signal. To most radiomen CQ meant an "attention" signal. Where did the letter combination come from? The French, sécurité, (safety or, as intended here, "pay attention"). Sometime later, the amateur origin of the abbreviation was changed to the phrase "seek you." [12]

In the spring of 1909 a young tall and thin 18-year-old graduated from Yonkers High School. Edwin Howard Armstrong was truly one who advanced the art of radio communications. Upon graduation he enrolled in Columbia University's School of Electrical Engineering. After school and in his attic at home, Armstrong was looking for a method of improving the received signal. He found that by feeding some of the plate-circuit signal back into the grid-circuit of vacuum tube he could increase the sensitivity. It amplified the signal a thousand times. In early 1913 Armstrong made a second discovery. If he increased the feed-back or regeneration, the circuit would oscillate and produce a continuous-wave signal.

Edwin Howard Armstrong graduated in June, 1913 and his father put up the money to pay for the patent and legal fees. His incorrect characterization of his invention delayed the process and he ended up with a flawed patent. Armstrong's mistake would trouble him for the rest of his

Edwin
Howard Armstrong

life. [13] Although not understanding the facts of electronics the Supreme Court after many years of litigation sided with DeForest. Upon graduation he accepted a job at Columbia College teaching classes in wireless. [14]

The International distress call went through a lot of changes in the last century. QRRR, was the first amateur distress call. There was a need for its use in the very first organized amateur emergency nets. This emergency signal was set up because of the frequent failure of the railroad telegraph landlines. The ARRL adopted QRR for a calling station to use when passing railroad traffic related to some emergency. Sometime later the call was changed back to QRRR because of a conflict in definitions with the international Q signal QRR.

In 1904 the Marconi Company coined one of the first distress calls as CQD. It was a combination "general call" (CQ) and the letter D for "distress". Many competing companies didn't want to follow the Marconi Companies system. In 1906 the problem was so bad that a brand new distress call was proposed.

The United States, Germany and the British all had different ideas concerning the distress signal. The Americans suggested NC which were already recognized in the International Signal Code for Visual Signaling. The Germans wanted SOE, which was already being used on their ships as a signal similar to CQ. The British wanted to stick to the Marconi signal CQD.

All of the delegates at the convention found SOE acceptable, except that the final E could easily be lost in the noise. That's when the letter S was substituted, making it SOS. The convention delegates all agreed that SOS should be sent as a single code character with a sound unlike any other character. Even though it was unanimously adopted, CQD remained in use for some years particularly aboard British ships.

In 1912, after the Titanic disaster, SOS became universal. The CQD signal gradually disappeared. Jack Phillips, the Titanic radio operator, sent both CQD and SOS as the ship was sinking. He was sure that there couldn't possibly be any misunderstanding. [15]

Chapter 3: Hiram Percy Maxim – The Early Years

On September 2, 1869 in Brooklyn, New York there were no telephones, electric lights, automobiles, power street cars or bicycles. That date was when Hiram Percy Maxim first saw the light of day. He was born into a house of inventors. Percy's father, Sir Hiram Stevens Maxim, had been knighted by Queen Victoria and invented the portable machine gun. Percy's uncle, Hudson Maxim, had created new explosives and propellants. In 1875 the Maxim family moved to a farm in Fanwood, New Jersey. After just two years they moved back to Brooklyn. [1]

Sir Hiram Stevens Maxim

Queen Victoria

In 1880 Percy showed a passion at a young age for mechanical things.

> One day I saw in Crandall's toy-store on Fulton Street a small stationary steam engine. It was a little bit of a thing, having a cooper boiler which would hold not more than a quarter of a teacup of water. It had a diminutive alcohol lamp under the boiler and a single oscillating engine on top of the boiler. It was a very primitive sort of a steam-engine, but it was real and it would run on steam. When my father came home that evening I told him of what I had seen.

> Some twenty-nine years after my father gave me this steam-train I gave my son one almost exactly like it. He got the same exquisite pleasure from his that I did from mine, and he did better than I did, for while he played regularly with his train for many years, it is still in good running order today, though it must be over twenty years old. [2]

After graduating from the Brooklyn School system in 1884, Percy attended MIT (Massachusetts Institute of Technology). At graduation in 1886 he was the youngest member of his class at age 16.

At the age of 17 Percy went to work as an electrical engineer at the Jenny Electric Company in Fort Wayne, Indiana. A year later he got another engineering job closer to his family, working for the W.S. Hill Electric Company in Boston. In 1888 at age 19 Percy became engineering superintendent at the American Projectile Company in Lynn, Massachusetts. This company would later become General Electric.

Percy by age 23 found his current situation frustrating. He had only two options to get to and from work, walking or bicycle. A horse required too much time and effort. Percy envisioned a horseless carriage. Without mass communications as we have today, he didn't know that many others were working on the same idea. Each night on his way home from work Percy's thoughts wandered. He thought of a small engine replacement for his legs pumping up and down running on gasoline. One night after the factory closed Percy tried one of his many experiments. He added a drop of gasoline inside a six-pound cartridge case followed by a lit match. This was followed by a violent explosion. Percy considered this a simple get-acquainted test.

If he did construct an engine, where would it go on the carriage? The terms carburetor, spark plug, magneto, dry cell, clutch or gears were not in the 1893 vocabulary. Percy moved forward by purchasing a used Columbia tandem tricycle for thirty dollars. The front and rear tires still had some rubber left. After many hours Percy designed a three-cylinder, four-cycle air-cooled apparatus. He left some of the minor details for a later date such as: muffler, manifold, carburetor and lubricating technique. Percy spent hours which turned into weeks in hopeless cranking with no results. By this time everyone in the factory knew what Percy was up to. By this time in his experimenting he had connected the engine shaft to a lathe. The problem appeared to be the gas-air mixture. One evening with some workers behind him, Percy decided to focus on the operation of the engine's needle valve. At one point the room filled with blue smoke followed by many loud explosions. Percy's assistant that night made it quickly to the door along with the others in the room. A month later with a muffler installed he tried the engine again. The engine ran for a very short time and then stopped. Percy concluded that without a load the engine was a runaway. In later years Percy said that this project was "a horrible example of how not to proceed."

After attaching the engine to the trike he had purchased, Percy attempted to ride it down a very steep hill. He attempted to pedal the trike which would eventually start the engine. It didn't happen until he tried a leaner mixture. Smoke, fire, gasoline and gravel from the road was everywhere. Percy was in a ditch, dazed with torn trousers, minor cuts and bruises with his wrecked tricycle. He pushed the trike slowly back to the factory. From this disaster Percy determined that he needed a clutch.

Percy on his trike - 1883

Horseless Carriage Days

During his vacation he visited the Pope Manufacturing Company, in Hartford, Connecticut. He still had clutches on his mind. Percy left at the end of the day with a job as head engineer in a new department of motor carriages. In July, 1885, Percy moved to Hartford, where his "troubles began."

Percy obtained a four-wheeled horse vehicle called a Crawford Runabout which he installed his engine from the tricycle. One early morning in August, 1885 Percy and his mechanic pushed the carriage out of the factory's back door. His goal in this early hour was to achieve a three percent grade up Park Street at fifteen miles an hour. Through strange noises and smells, Percy did it. A few days later he followed that up with a drive into the country and back, a distance of about five to six miles each way. October, 1885, Percy became the driver in Connecticut's first motor car. [3]

Mister Pope

Percy's tricycle with its gasoline combustion engine was considered by most historians as one of the first vehicles of its kind. He continued to invented several different types of automobiles with all types of engine. There were electric engines, gasoline engines, built into three wheel and four wheel vehicles. Just before 1895 ended Percy won the Times-Herald race with one of his gasoline carriages. Three years later with the first four-wheeled gasoline auto in Hartford Percy won the race to Boston and back. He was not very popular with the steam engine drivers in the race that usually won. With several races behind him, Percy went on to perfect his Maxim Mark 8. This model took the form of a real automobile. It had an engine under the seat, four wheels, a clutch mechanism and a crank start. Up till now autos used a fly-wheel spinner to start the engine. Percy was considered by many as an absolute genius. [4]

Hiram Percy Maxim entered and won all of the races before turning to a new project. He wanted to design and build the Mark 1 electric horseless carriage. Percy wanted a vehicle that was reliable, with no noise, or foul smells. After finishing the Mark 1 he put on a public demonstration which was a complete success. During the construction of the vehicle Percy kept a very accurate journal. He did this with all of his projects.

"Mark I memo... for 10 mile speed, fall in voltage on present #2 batteries makes it impossible to calculate what induction is necessary to get torque and speed to correspond. Mark I weight, 1550 pounds (with passengers 1825), sample test run... mileage 9.75, time 55 minutes... 10.63 miles per hour, stops included." [5]

Percy worked with Hayden Eames of the Pope Manufacturing Company in the actual design of the electric carriage. During the design period Percy made several trips to the Electric Storage Battery Company in Philadelphia. On his first visit Mr. Gibbs, the president, told Percy he thought it was a scatter-brain idea to put storage batteries in horseless carriages. Percy convinced him that he was serious. During one of their later discussions, Percy mentioned that the batteries were just a bit too heavy. Gibbs answer was given with a smile. "You must realize, Mr. Maxim, that the good Lord never intended a storage battery to be made of anything but lead." They compromised on the battery which was lighter which would limit mileage and speed somewhat. [6]

Houseless Carriage Days
Hayden Eames

Needless to say, the Mark I was a success. It was followed by the Mark II. a two-cycle project which turned out to be a miserable failure. Instead of being fired, Percy was put in charge of a project to produce a small gas trike to aid merchants in delivering packages. The Mark VII was ready for testing in February, 1897. During all of the new model carriage testing and troubleshooting those that had been sold, Percy was courting a young lady. A mere fifteen weeks after Percy's 29th birthday, December 21, 1898, he married Josephine Hamilton. She was the daughter of a former Maryland governor. The couple moved to Hartford, Connecticut. In October, 1900, a son was born, Hiram Hamilton Maxim. [7]

19

Horseless Carriage Days

The Mark VII built in 1897

In 1899, at a small racetrack in Branford, Connecticut, Percy beat a Stanley Steamer with his Mark VIII machines. Percy felt that this was the first motor track race in America and he, its first winner. Thirty years later at a product meeting he met the man he beat in the Stanley Steamer. [8]

Percy continued to design the Mark XII through the Mark XIX. He also continued to test carriages, both gas and electric until 1901. That's when he considered it to be the end of the Horseless-Carriage Days. Negotiations had ended in 1900 between the Pope Manufacturing Company and New York interests on forming the Columbia Automobile Company.

The turn of the 20th Century was observed by many historians as a era of revolutionary changes in transportations and communications. Percy was soon to be leaving one of these categories and entering the other. He became chief engineer at the Electric Vehicle Company in 1903. On July 4, 1906 a daughter, Percy, was born in the Maxim family. That same year Percy went into business with the best man at his wedding, Thaddeus Goodridge. Their venture was to produce an electric automobile. During Percy's research he

Percy with rifle silencer in 1909

had been thinking about his father's invention of the machine gun. A muffler for a gas engine and a muffler for a firearm merged in his mind. The idea of a silencer came to him in the bathroom while watching water go down the drain. Percy built his first silencer and applied for a patent in 1908. The primary purpose of this silencer was its use by the U.S. Army.

A Genius in the Family
Sir Hiram Stevens Maxim and grandson, age 8

While still working with Goodridge, Percy met and hired a young man to work for him. They developed a lifelong relationship. His name was Roland Bourne and his hobby was amateur radio. [9]

Chapter 4: Hiram Percy Maxim and the ARRL

One story has it that in 1911 Percy became an amateur radio operator because he was influenced by Roland Bourne his longtime friend. His son, Hamilton at age eleven also got his license. Percy's first station was SNY, then 1WH and 1ZM. His station had a range of one city block. Another story circulating was that Percy's son, Hamilton saw a radio receiver in a toy store and told his father. Percy went with him to the store and bought one. That story has not been validated, the first one will have to do.

Within a short time Percy had improved his coverage to five miles. Percy continued to work on his station. His attention was divided between his signal coverage and the Radio Act of 1912. Percy said of the act, "one of the most constructive and valuable bits of legislation that a Congress has ever enacted." Amateur Radio got a boost seven months after the Act was signed. In March, 1913, a group of amateurs provided aid in a destructive windstorm in Ohio and Michigan.

January, 1914, was the month that began a personal achievement for Percy and a historic journey for amateur radio. It was the first meeting of the Radio Club of Hartford. Percy presided and before the meeting was over, a "blue-eyed boy of eighteen," was selected as secretary. His name was Clarence Tuska.

Hiram Percy Maxim
Percy in 1911 wearing an arm band in mouring for the death of his mother.

Many things in the past concerning the automobile seem to just pop into Percy's head. His thought process continued with his amateur radio hobby. [1]

At that first meeting David L. Moore was elected president of the club. It was decided to have bi-monthly meetings. The first order of business was to create a club constitution which was adopted at the next meeting. There were twenty-three charter members which grew to thirty-five by the March 9th meeting. Sometime between March 9th and April 6th meeting the wheels were set in motion for amateur radio history.

Percy: Radio now had a possible range of about one hundred miles using his one-kilowatt station. In March, Percy had been unsuccessful in

Clarence D. Tuska

purchasing an Audion (vacuum tube). He heard that an amateur in Springfield, Massachusetts, thirty miles of Hartford had one. After several attempts Percy couldn't "raise" Springfield. After mulling over the problem for a few minutes,

he came up with the solution. A young man, Windsor Locks, had come to the club meeting and he lived in between Hartford and Springfield. So, why not have Locks relay a message concerning the vacuum tube he needed? Percy knew that ships do it, so why can't amateurs also. He even came up with a possible name, "American Radio Relay League." Percy didn't want his idea to get too cold so he presented it at the next meeting of the Hartford Club.

Message handling – for pleasure, for friends, in time of emergency – was rapidly becoming the predominant theme in amateur radio. It was shortly to assume a position of dominant importance in the development of an amateur organization of truly national character.

Application for Membership in the Radio League of America

I, THE UNDERSIGNED, a Radio Amateur, am the owner of a Wireless Station described in full on the back of this application. My station has been in use since.............................., and I herewith desire to apply for membership in the RADIO LEAGUE OF AMERICA. I have read all the rules of the LEAGUE, and I hereby give my word of honor to abide by all the rules, and I particularly pledge my station to the United States Government in the event of war, if such occasion should arise.

I understand that this blank with my signature will be sent to the United States Government officials at Washington, who will make a record of my station.

Witnesses to signature: Name...

.... City...

State..

.... Date..............................191

Describe the apparatus of your station on the back of this application.

In the event of national peril, you will volunteer your services as a radio operator in the interest of the U. S. Government?.............

This last question need not be answered unless you so desire it.

The club membership voted unanimously to adopt Percy's name and function. Application forms were printed. Percy and Clarence wrote letters to every amateur radio operator they could think of conveying the formation of the American Radio Relay League. Membership was free and there were no dues. Postage was contributed by both Percy and Clarence. Response was overwhelming and by late June successful relays had been made from Hartford to Buffalo and Boston to Denver. Nearly 200 stations had been established by August, 1914. Coverage was increasing from Maine to Minnesota and from Seattle to Idaho. In September the league published the locations of 237 stations in thirty-two states and Canada. The following month they published a *List of Amateur Stations* which today we would recognize as the first Call Book. [2]

An Amateur Wireless Magazine
PRICE 10 CENTS
QST
DE
The American Radio Relay League
FOR THE MONTH OF JANUARY
NINETEEN HUNDRED AND SIXTEEN

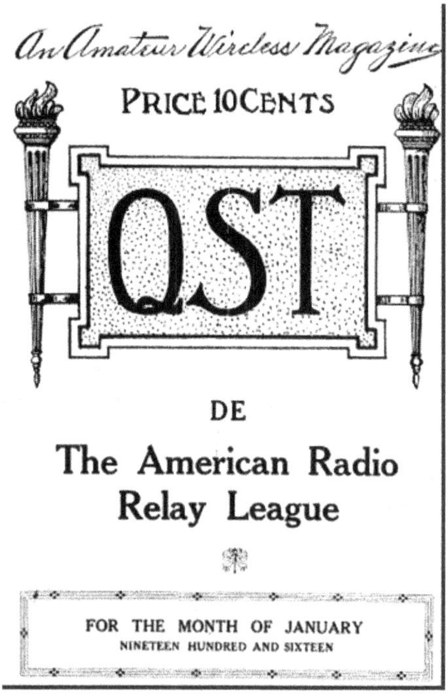

As Percy and Clarence were moving forward with the Relay League's agenda, Edwin Howard Armstrong received his receiver patent on October 6, 1914. His regenerative circuit became very popular. Armstrong gave a report to the Institute of Electrical Engineers in 1915 on the operation of the three-element vacuum tube. This was the first time a correct explanation was given. It contradicted DeForest's assumption that no alternating current was present in the tube's plate circuit. [3]

Percy wanted his organization to get some recognition. So, late in 1914 he took a trip to Washington, D.C. to spread the word. One of his stops was with the Commissioner of Navigation of the Department of Commerce. Percy also wanted to get special licenses for specific relaying stations.

February, 1915 was a milestone. The League was growing by leaps and bounds but lacked the necessary funding. With no dues, there was no money coming in for pamphlets, forms and postage. Percy separated himself and Clarence from the Harford Club and incorporated the American Radio Relay League. By putting their heads together, the magazine, QST was born. The first issue came out in December, 1915 and was sold by subscription only. The cover was designed by Clarence's uncle and contained 24 pages. A three-month trial subscription cost twenty-five cents. Clarence Tuska became the magazine's first business manager as a student in college. [4]

ANNOUNCEMENT

¶ Q S T is published by and at the expense of Hiram Percy Maxim and Clarence D. Tuska.

¶ Its object is to help maintain the organization of the American Radio Relay League and to keep the Amateur Wireless Operators of the country in constant touch with each other.   .

¶ Every Amateur will help himself and help his fellows by sending in 25 cents for a three months' trial subscription.

THE PUBLISHERS OF Q S T

The ARRL membership jumped from 635 in December 1, 1915 to 961 on January 10, 1916. [5] There was one person who wasn't that happy about the membership increase. Hugo Gernback's magazine, *The Electrical Experimenter* in 1916 refused advertising from the ARRL, as a competitive threat to his organization, Radio League of America. Gernback was what you would call an entrepreneur today. He invented the first walkie-talkie in 1909 and worked with early television in 1928. Gernback coined the phrase, "science fiction" and published its first magazine. His followers consider him the father of that genre. The award for science fiction is called the "Hugo," named in his honor. Gernback died in 1967. [6]

Percy, always looking for improving the League, had planned to implement six-trunk lines which would cover the United States. On Washington's Birthday, 1916, Percy held the first country-wide relay which involved state governor's and President Wilson.

1916 was the same year that Percy's father Sir Hiram Stevens Maxim died at seventy-six in England. He was a famous and rich man, but as Percy put it, "He never quite learned how to be a father."

One of the reasons that QST grew in popularity was due to T.O.M., or The Old Man. Only six people knew that the author of the humorous column was Percy himself. Percy poked fun at nearly every aspect of amateur radio.

Another big day, January 17, 1917, for Percy and the ARRL. It was a transcontinental relay. This was followed on February 6th with an East Coast to West Coast relay. A relay was made across the U.S. complete with the originating station receiving an answer. Start to finish was one hour and twenty minutes. Before the end of February the ARRL had put in place a governing structure, adopting a constitution and developed policies. [7]

A letter was mailed to every amateur radio operator in April, 1917. **The United States of America had entered World War I.**

"To all Radio Experimenters,
"Sirs:
"By virtue of the authority given the President of the United States by an Act of Congress, approved August 13, 1912, entitled, 'An Act to Regulate Radio Communication,' and all other authority vested in him, and in pursuance of an order issued by the President of the United States, I hereby direct the immediate closing of all stations for radio communications both transmitting and receiving, owner and operated by you. In order to fully carry this order into effect, I direct that all the antenna and all aerial wires be immediately lowered to the ground and that all radio apparatus both for transmitting and receiving be disconnected from both the antenna and ground circuits and that it otherwise be rendered inoperative for both transmitting and receiving any radio messages or signals, and that it so remain until this order is revoked. Immediate compliance with this order is insisted upon and will be strictly enforced. Please report on the enclosed blank your compliance with this order; a failure to return such blanks promptly will lead to rigid investigation."
"Lieutenant, U.S. Navy, District Communication Superintendent." [8]

Shortly after the announcement was made to amateurs, the House of Representatives in Washington introduced a bill. This bill proposed that all radio communications, including amateur, should be handed over to the Navy. Percy was in the Capitol in a flash lobbying against the proposed. The bill died in committee.

The U.S. Military establishment found themselves in a pickle. There was a complete shortage of radio operators and radio instructors. The military contacted Percy. They needed five hundred qualified radio operators now! They also wondered if amateur radio equipment could be modified for military operation. Before the end of World War I, four thousand radio operators had seen duty. But, the amateur radio ban was not ended.

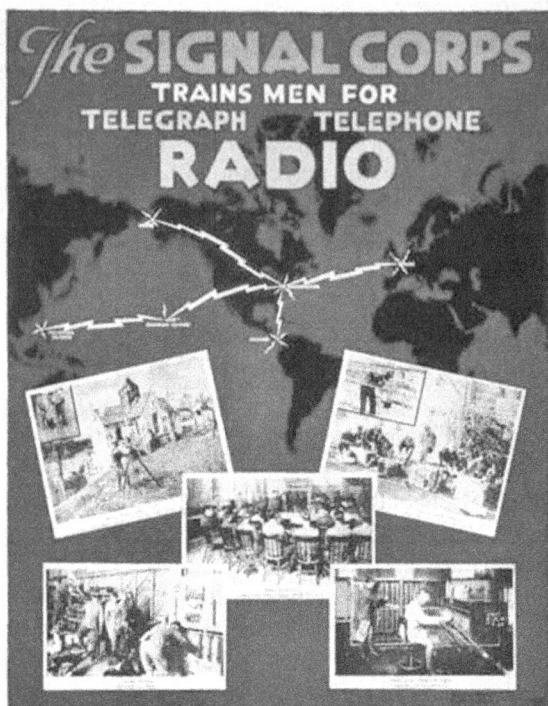

There is in Your Locality a
U. S. ARMY RECRUITING OFFICE

Clarence Tuska went into the service. Percy wrote to him on October 20, 1917, November 15, 1917 and March 31, 1918 with cheerful and humor to keep his spirits high. He did write about his sadness concerning the amateur radio ban. [9]

Chapter 5: War over – Back on the Air

The war was over and the important part played by amateur radio in the war was explained by Lieutenant Clarence D. Tuska. The ARRL's secretary who volunteered for service was put in charge of radio training in the Air Service. He had this to say:

> "The amateurs have come across in the case of the Army. . . I have turned out a whole lot of operators for the Air Service and have pretty well acquainted with the type of human it takes to make a first-class operator. . . The very first sort of a student we looked for was the ex-amateur. He seemed to have had all the experience and all we have to do is acquaint him with a few special facts and he is ready for his Army job. If we can't get an amateur or a commercial radio operator, then we try to convert a Morse (wire) operator, but it's a pretty hard job. After the Morse man, we take electrical engineers, and from them on, but a man without previous experience is almost hopeless as far as my experience has shown. Of course we can make an operator of him in fifteen or sixteen weeks; whereas, the other way an amateur is fitted in as few as one hundred hours. They've surely done their bit and I am mighty proud I was one." [1]

While the war was still raging, Edwin Howard Armstrong was working on a project concerning the reception of shortwave signals. The result of his undertaking was an eight-tube receiver in the summer of 1918. It had higher stability, improved selectivity and could amplify higher frequencies. Armstrong called it a superheterodyne. He applied for a patent on February 19, 1919 which was granted sixteen months later. [2]

November 11, 1918 celebrated the armistices, but the war on amateurs was far from over. Another attempt was made in Washington to put the Secretary of the Navy in charge of all radio in the U.S. Percy, at his own expense went to Washington again to attend the hearing on the bill. Again, the bill died in committee. [3]

Edwin Howard Armstrong found himself at "war" in the U.S. He received word that DeForest was using his regenerative circuit ignoring Armstrong's patent. In litigation Armstrong ran out of funds and reluctantly sold his regenerative receiver patent to Westinghouse for $100,000. [4]

The League met in March, 1919 with twenty-three dollars in the treasury. A vote was taken to resume QST which was halted when the war started in 1917. The eleven board members present each put money in a hat to publish the new QST issue. [5] Clarence D. Tuska stepped down as ARRL Secretary at this meeting. This he felt his interest in the radio manufacturing industry would be a conflict of interest. [6]

Hiram Percy Maxim Jan. 2, 1919

THE MAXIM SILENCER CO.

HARTFORD,

HIRAM PERCY MAXIM.   CONN.
ENGINEER

The Navy Department announced on April 12, 1919 that the ban on receiving was lifted. Until the President declared peace, transmitting was still banned. Percy urged the ban be lifted and got the aid of the Assistant Secretary, Franklin Delano Roosevelt. The Director of the Navy Communication Service lifted the ban on September 26, 1919.

Returning to the air was a slow process. While the Navy had been in control for two and a half years, all radio amateur licenses had expired! Temporary verification got amateurs back on the air by November, 1919. A month later, Percy took part in the first post-war transcontinental relay using his new call 1AW. [7]

A new word popped up in 1919. Mayday became another distress call that has nothing to do with the first day of May. Rumor has it that the American aviators in World War I picked this term up from the French. In doing so they mispronounced the word. The original French word, "m'aidez" means "help me". It is easy to see the mistake that has been around since the First World War. [8]

The Radio Act of 1912 ushered in an era of mandatory licensing. It created the Amateur First Grade and Amateur Second Grade operator license. The two licenses were identical except for one small item. If you could not attend an examination session, you simply sent a letter stating that you could meet the requirements of type two license. That got you a Second Grade license. In 1919, along with the written test, the code requirement was increased from 5 to 10 words per minute. [9] Amateurs added another item to their operating procedure, an acknowledgement card, or QSL card. The earliest account of sending a QSL card was in 1916. Station 8VX in Buffalo, New York sent one to 3TQ in Philadelphia, Pennsylvania. The standardized card as we know it today was created in 1919 by C.D. Hoffman, 8UX, in Akron, Ohio. [10]

By now you have figured out that Percy liked a challenge no matter what its source. In 1920 he made a bet with his sister-in-law that he could write a movie script and have it produced. Percy felt that anyone could write a scenario. Percy was surprised when the script was accepted and on its way to a Hollywood production.

Under the title "A Virgin Paradise," a Fox Photoplay Production, it had its premier at the Park Theater in Hartford. The newspaper account referred to it as "a story of the jungle and civilized hypocrisy. . .

From her popularity in *The Perils of Pauline*, Pearl White was chosen for the leading role. Local newspapers had a heyday with Percy's new project.

Pearl White

A newspaper headline commented, Maxim, who took the noise out of bullets, put pep into films" – giving Pearl White "an opportunity to frolic with lions and monkeys."

Percy's two mast antenna system appeared on the cover of the July, 1920, QST which also displayed new diamond shaped logo. [11]

In early 1921 the Radio Corporation of America made an announcement that would have lasting implications. They were making a series of transmitting tubes available to amateurs.

The first, the historic UV-202, was a 5 watt tube priced at $8. An equivalent tube can be purchased today for 69 cents. Yet to the amateur of 1921 the UV-202 was the answer to a long and weary prayer – and he was glad to have it at any price. In middle March, R.C.A. promised to put its UV-203, a 50-watt tube costing $30, on the market; and in April, the biggest bottle of all, the G.E. pliotron, the successor of the aristocratic old "P" tube, a 250-watter costing $110.

ARRL Official Logo

Radio amateurs couldn't seem to stay out of hot water. With a total of about twelve hundred broadcasting stations in operation, many commercial users were experiencing interference from amateurs. Secretary of Commerce Herbert Hoover formed the First National Radio Conference in Washington, from

February 27th to March 2, 1922. Secretary Hoover was on the side of the amateurs apparent by his remarks to the press:

"I would like to say at once that anyone starting any such suggestion that this conference purpose or had any notion of limiting the area of amateur work was simply fabricating. There has never been any suggestion of the kind, never any discussion of the subject in any shape or form. The amateurs were asked to be represented in the conference and they are represented here today, and the starting of that sort of information is one of the most treacherous things that can be done. So I wish to sit on that right from the start – that the whole sense of this conference has been to protect and encourage the amateur in every possible direction."

Herbert Hoover

The very first time the amateur radio operator was given recognition and an increase in the number of bands in which they may operate their equipment. This conference recommended that the region between 150 meters and 275 meters be used by amateurs. With one exception, that the area between 200 meters and 275 meters be shared with technical and training schools.

"An amateur is one who operates a radio station, transmitting or receiving, or both without pay or commercial gain, merely for personal interest or in connection with an organization of like interest." [12]

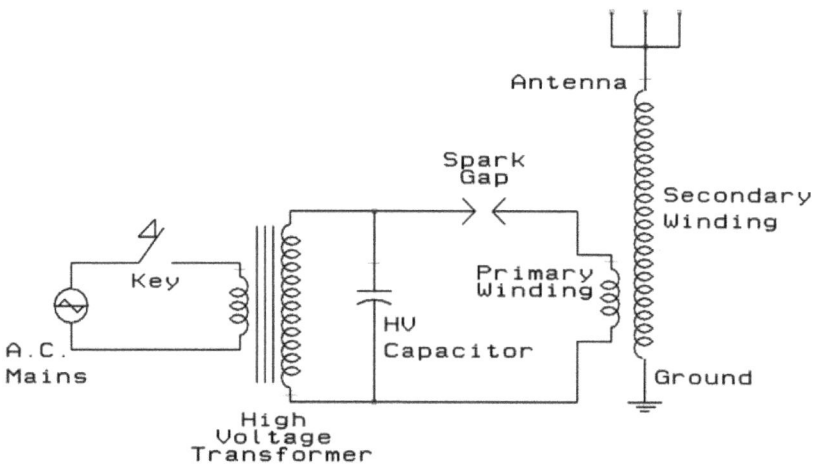

ROTARY SPARK GAP

A **Rotary Spark Gap** is required in every transmitting station by the Federal authorities, for the reason that this type of gap produces a pure wave of low damping decrement. It also increases the efficiency of any transmitting station from 20 to 30 per cent.

This Rotary Spark Gap emits a high musical note, more audible to the human ear, can be heard at greater distances than the note from the stationary type, and cannot be mistaken for static or other atmospheric disturbances, a fault common with the stationary gap due to its low frequency note.

The rotating member has twelve sparking points mounted on a hard rubber disk and is carried on the motor shaft.

Also fitted with two stationary electrodes with special adjusting devices.

The Gap can be successfully used on any of our spark coils or transformers up to and including 1 K. W. capacity.

Our standard Globe Motor is used, which will operate on 110 A. C. or D. C. circuits and attains a speed of 4,500 R.P.M. Also made with our Globe Battery Motor, which can be operated on a six-volt circuit.

List. No.		Price
222	Mesco Rotary Spark Gap, 6 volt	$12.00
223	Mesco Rotary Spark Gap, 110 v., A. C. or D. C.	13.00
216	Rotary Unit only, with two Stationary Electrodes, 1 3/16 in. shaft	5.00

QST Magazine January 1916

The manufacturing of new vacuum tubes was good news to many amateurs and bad news for others. It was the beginning of another conflict. This time it wasn't between the U.S. Navy. It was amateur to amateur, "spark" vs. "CW!" Before WWI the commercial and government radio stations worked great distances with CW. However, it was common knowledge that during the three years after the war, no amateur on the East coast ever had his signal heard in Europe. Those dyed-in-the-wool operators held-on to their romance with spark. They would rather give up amateur radio that switch to CW. To them, each of the CW signals sounded the same.

One way you could see the slow death of spark was the pages of ads in QST and other periodicals starting in 1922. Since most of the ads appeared for months on end, it was apparent their owners couldn't even give that equipment away.

Here is an example of humor as used in the September, 1923, issue of QST magazine, written by Porter Bennett, 5IP:

> "Here it is, OM. The idea hit me while listening in late night. Nary a spark did I hear, and I thought how good it would be to hear one closing down with the power still on and the note descending 'dah-de-dah.' But all I could hear were signals that stopped with a sudden abruptness that left something lacking. C.W. is better than spark and I like it better, but the ringing of the cowbells sounds sweeter to a farmer still."
>
> *"Spark is dead!"* [13]

The year 1923 would begin an exploration into amateur radio activity, more than previous years put together. Vacuum tubes would be the most significant item to open the potential of amateur radio. CW would begin to spread as spark began

33

to fade. Many spark fans displayed "Spark Forever" on their QSL cards.

Amateur radio had an opportunity to prove its worth, which it did with flying colors. A complete amateur station was donated by the Zenith Corporation to be put aboard a ship bound for an Arctic Expedition. Donald H. Mix from the ARRL went along as the ship's (amateur) radio operator. He sent back a weekly 500- word message to the North American Newspaper Alliance. Captain MacMillan said that "no polar expedition will attempt to go North again without radio equipment."

Captain MacMillan and Percy

The year 1924 kept Percy very busy. The ARRL asked him to represent them at a conference in Paris on March 24th. Then on June 14th, Percy gave the commencement address at Colgate University. Radio was the substance of his speech. By winter when a storm just paralyzed half of the U.S., Percy, the ARRL and amateurs came to the rescue with emergency communications. [14]

ARRL

Percy at his rig - 1924

34

American Radio Relay League Station 1AW

Hiram Percy Maxim, Owner   276 No. Whitney Street, Hartford, Conn.

Radio *U - 9C4*   *1-29-25*

Your............................signals were ^worked_heard here on....................at about

............A.M. E.S.T.
............P.M.
on Tuska three-cir-
cuit tuner and two
stages audio amplifi-
cation, Baldwin
phones.
Audibility ....

1AW

Wave-length ....
QRM ....
QRN ....
QSS ....
Weather ....
Tone ....
Wave ....

Remarks: *Thie QST a very card I put on 83 meters and have lots of punch so I ought to be getting me.*

Would like report on 1AW's signals if you hear them. Best 73's.

Hiram Percy Maxim Operator

ARRL

Percy's QSL Card

After a year's negotiation on the part of ARRL, the Commissioner of Navigation of the Department of Commerce endorsed a plan for a new amateur license. It would permit using 75 to 80 meters, 40 to 43 meters, 20 to 23 meters and 4 to 5 meters for amateur operation. It was also stated that spark would no longer be permitted on these new bands. However, amateurs could not use these bands with their existing licenses. [15]

The First International Amateur Radio Congress opened on April 14, 1925. Twenty-five nations were represented. On the fourth day the delegates elected Percy as its international president. Back from Europe Percy acquired a summer home near Lynne, Connecticut. During his year friendly relations had finally been established between the military and the ARRL. On June 9th, Percy was commissioned a Lieutenant Commander in the United States Naval Reserve. In August the U.S. Army promoted the establishment of the Army-Amateur Radio System. The remaining part of the year Percy worked on various projects.

Percy started 1926 with a social event. Clarence Tuska created the Tuska Radio Company with Roland Bourne as company engineer and patent director. Percy and his wife took his boat, the Seagull, on a European tour. He spent a lot of time with photography. Percy brought a crank camera complete with tripod. In this year he founded the Amateur Cinema League which was a cross between business and pleasure.

All good things must come to an end. In the summer the wheels fell off of radio operations. On April 26, 1926 the decision of the Zenith case made the radio regulations null and void. The Attorney General announced that the federal government had no control over radio that does not appear in the 1912 Act. At that time high-frequency allocations and radio broadcasting were non-existent. [16]

36

Chapter 6: Amateur Radio – A New Beginning

Commercial broadcasting stations were mainly involved in the "Summer of Anarchy" by moving to any frequency they desired. While amateur radio operators stayed within their allocated bands.

The year of 1926 also brought crystal-control to amateur which added needed stability that had been needed for years.

In order to solve the problem once and for all Congress created the Federal Radio Commission (FRC) through the Radio Act of 1927. It was the first time that the word "amateur" appeared in a federal document. The Act was signed into law on February 18, 1927.

> The purpose of the Radio Act of 1912 was to provide for the licensing of operators, the lettering of stations, the minimizing of interference, the facilitating of radio communication and particularly to give the government the right of way to distress or danger signals or other important intelligence. The language of the act, its general scope, and the nature of the subject regulated shows that Congress did not intend to give to the Secretary of Commerce and Labor any discretion in the issuing of licenses. The supervision and control was taken by Congress upon itself leaving the Secretary of Commerce a mere authority to deal with the matter as provided in the act and giving him no general regulative power. The duty, therefore, of naming a wave length was mandatory on him. His only discretionary act was in selecting the wave length within the limitations prescribed by the statute which in his judgment would result in the least possible interference. The issuing of a license was not dependent on the fixing of the wave length. The wave length named by the secretary merely measured the extent of the privilege granted to the licensee.2 Mr. Hoover acting under this statute has played an important part. Instead of prematurely recommending necessarily inadequate laws for this young and growing industry he has wisely been content to call conferences of engineers, scientists, amateurs, broadcasters, manufacturers, distributors, and others interested in the trade, thus keeping abreast with the art and preventing himself from acquiring any frozen views on the subject. The result of. his policy is the Radio Act of 1927 just enacted by Congress.

At the moment when President Coolidge signed the radio bill there were in existence in the United States 733 program stations, 14,768 amateur stations, *22* trans-ocearfic stations, 63 coastal general service stations, 74 point-to-point limited service stations, *207* limited commercial stations, 179 experimental stations, 38 technical and training stations, and **2,035** ships equipped with radio-a total of **28,119** stations. [1]

Calvin Coolidge

In the fall of 1927 yet another conference was held in Washington D.C. The International Radiotelegraph Conference hosted seventy-four nations represented by 351 delegates. Percy had concluded that most of the delegates had no understanding of amateur radio other than just an annoyance to the airwaves. Percy and the ARRL vice-present used diplomacy on all the delegates they spoke with. The conference outcome was mixed. American amateurs lost over thirty-seven percent of their bandwidth, but gained more privileges. The gain was ten meters. With wavelength compromises, International call signs were also assigned.

The year 1928 started out more calm than the year before. Percy was serving a second term on Hartford's Board of Health when he was elected to the Harford Board of Education. Percy was also selected to serve as Aviation Commissioner for the city of Hartford. All of this was happening while he was still involved in his silencer business and opened new ventures. Percy called his new unit, Maxim Evaporators, a low cost way of converting salt water into fresh water. Still on the go, he and his wife took a Mediterranean cruise.

Percy's Silencer Company was hurt just like most business during the Depression of 1929. To add to his financial loss, his wife Josephine had a mild stroke. [2] On March 1, 1929, the Army-Amateur Radio System had a newly revised plan. In times of emergency and distress, the Army and Red Cross would work together.

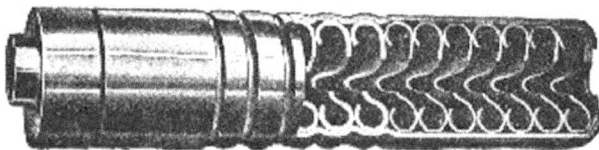

Percy Maxim Silencer

It was back to Washington D.C. in the first month of 1930, with a bill introduced in the U.S. Senate by Senator Couzens of Michigan. The bill would create a national communications commission to control all wire and wireless communications. Percy testified at great length on January 31, 1930 in front of the Interstate Commerce Committee. He passionately spoke about the importance of amateur radio. The bill was defeated. On April 5, 1930 the Federal Radio Commission did make some basic changes to amateur regulations. These changes involved adequately filtered DC power supplies, excessive modulation, changing the 28 and 56 megacycle bands from shared to amateur exclusive, the keeping of logs and the definition of "quiet hours." [3]

Even though there was a depression, amateur radio enjoyed its greatest growth. From 16,829 to 46,850 amateurs were licensed. [4] An amateur could have built a station in 1934 for $50 that would have cost three times that amount in 1929. Also equipment such as the superhetrodyne receiver was very effective. It had taken more than ten years for amateurs to see the affects of Armstrong's invention. Amateurs were also moving to the five-meter band. [5] CW was still king, but some phone operations were being heard on the HF bands. Crystal controlled equipment was the norm. [6]

Percy continued to work on improving his inventions. The Hartford Courant newspaper carried a story in its October 30, 1930 issue. It describes the Maxim Window Silencer as "a boon to hospitals and office buildings." Once again the Hartford newspaper carried a story about inventor Maxim. It read "a plain metal box which keeps out all noise while permitting fresh air to enter." That device was breaking new ground in the field of air conditioning. Percy said, "The result of twelve years of work."

The Office of the Chief of Operations of the United States Naval Department sent a request to Percy on October 1, 1931. They asked him for an autographed photograph of himself. They wanted to hang his picture next to Marconi, Bell, Faraday and Morse. Later that month Percy received an invitation from the President of the United States to attend the dedication.

In the December, 1931 QST Magazine an editorial was published explaining the origin of the word 'ham.' According to the article it came from the British Cockney pronunciation of the word amateur (h'amateur), abbreviated into 'ham.' [7]

Yet another conference was held, this time in Madrid on December 9, 1932. There were nearly a hundred international associations and seventy-seven nations in attendance. No changes to amateur radio were made. The ARRL had representatives there in case amateur rules changes came up. This Madrid Treaty became law on January 1, 1934 and ended January 1, 1936. [8]

About this time Percy started writing his two autobiographical books about his early life. His first book was, *Horseless Carriage Days* and the second *A Genius in the Family*. Percy's second book was made into a movie called "So Goes My Love," starring Myrna Loy and Don Ameche. Percy did more writing about another one of his hobbies, astronomy. It was published in the Scientific American in the April, 1932, issue. He got interested in the planet Mars and by 1933 began his third book, *Life's Place in the Cosmos.*

Even though the economy wasn't improving, the activities of amateur radio were. The year 1933 saw a large increase in expeditions and exploration. This was also the year that amateurs received much credit when helping in disasters across the globe. [9]

Behind all of the amateur radio activity certain hams were trying to improve the hobby. Robert M. Moore, W6DEI, was a true pioneer. He published three articles on "single sideband suppressed carrier" (SSSC) radio transmission in a new magazine, *R9* in 1933 and 1934. Today we know it as "Single-Sideband," (SSB). This new mode should have "caught on" back in 1933. However it didn't for two reasons. First, the amateurs had just recently been introduced to amplitude modulation (AM) and didn't want to purchase more transmitting equipment. [10]

Another inventor that would have an impact on amateur radio a few decades into the future was Edwin Howard Armstrong. He was searching for a static free radio system other than the current amplitude modulated (AM) signal. Armstrong created a frequency modulated (FM) system and received a patent on his wide-band FM on December 26, 1933.

In 1934, President David Sarnoff hired Armstrong to come to work at RCA. Sarnoff met Armstrong in 1920 and was impressed with Armstrong's FM system, but also understood that it wasn't compatible with his AM empire. Sarnoff considered FM as a threat and refused to support it. From his laboratory constructed by RCA on the 85th floor of the Empire State Building, Armstrong conducted the first large scale field tests. This occurred from May 1934 until October 1935. The results; conclusion held that FM had far superior sound over AM.

David Sarnoff

In order to prevent FM radios from becoming dominant, RCA lobbied for a law from the FCC. It did not happen. So, a patent fight between RCA and Armstrong followed when RCA claimed they invented FM radio. RCA's historic victory in the courts left Armstrong unable to claim royalties on any FM receivers, including televisions, which were sold in the United States. He was left with no funds to fight on. [11] RCA also crushed Philo T. Farnsworth the true inventor of television.

Over the years, after World War II many FM equipment manufactures ignored Armstrong's patents. On January 30, 1954, Armstrong broken and in poor health jumped out of the window from his thirteenth-story New York City apartment.

In the spring of 1968 the last of his 20 FM patent cases was settled. His widow received $10 million. [12]

The passage of the Communication Act of 1934 was signed on June 19th by Franklin D. Roosevelt. It created the Federal Communications Commission which replaced the Federal Radio Commission. One item affected the amateurs which just reorganized the license structure into Class A, Class B, and Class C. During this year Percy wrote more newspaper articles and played with new color film for his movie camera. He used it during the Christmas holiday on family, especially grandchildren. This would be his and his wife Jennifer's last Christmas.

F.D.R.

Percy had planned a trip in January, 1936. He wanted to visit the Percival Lowell Observatory in Flagstaff, Arizona. He and Jennifer left by train on February 8th for a month's vacation. Percy became ill and was treated by a doctor in Kansas City. His symptoms deteriorated and he was taken to a hospital in LaJunta, Colorado. His family was sent for in Connecticut. Percy was delirious and never regained consciousness. On February 17, 1936, at age sixty-six, Hiram Percy Maxim died. The amateur radio world had lost its champion.

The dedicatory preface in Clinton B. DeSoto's book, *200 Meters and Down*, was written by Herbert Hoover in September, 1936:

> I well remember the battle in which Hiram Percy Maxim joined with me as Secretary of Commerce in setting apart definitely and for all time certain segments in the radio range and dedicating them for the perpetual use of the amateurs. The commercial value of these wavelengths was well recognized at the time and great pressures were brought to bear to allot them to commercial use. Mr. Maxim's sturdy mobilization of the thousands of amateurs contributed greatly to saving this field, which has now extended into world-wide use.

> The amateurs have performed many signal acts of public service not alone in the field of experiment and research but in the actual transmission of vital messages. Their art has added to the joy of life to latterly hundreds of thousands of men, women, boys and girls over the whole nation. Their international communications have a value in bringing a better spirit into the world.

41

I considerate it an honor to join in any tribute to the memory of Hiram Percy Maxim. [13]

Percy's long time friend, Clarence D. Tuska remembered him a few years before the ARRL in a book he wrote in 1957.

Maxim had both imagination and foresight – they are a kind of Siamese twins when we consider inventions and inventing. A few years after I met him he gave further evidence of foresight in suggesting that airplanes (aeroplanes then) be propelled like skyrockets. He patiently tried to explain that rocket propulsion had advantages over propeller drives. I was not old enough nor intelligent enough to appreciate fully his conception of what has now become the jet plane. [14]

Percy left the ARRL with a wonderful foundation. Apparent today, since the ARRL is still going strong after 100 years.

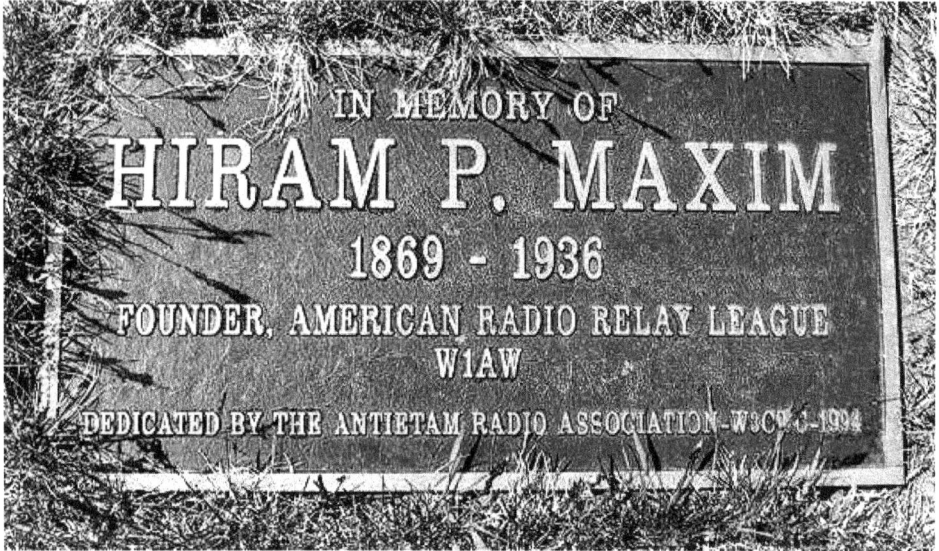

IN MEMORY OF
HIRAM P. MAXIM
1869 - 1936
FOUNDER, AMERICAN RADIO RELAY LEAGUE
W1AW
DEDICATED BY THE ANTIETAM RADIO ASSOCIATION-W3CWC-1994

Chapter 7: Clarence D. Tuska

This story isn't complete until the life of the ARRL co-founder has been told. Part of this may be a duplication of material but this material adds closure to a piece of history.

On August 15, 1896, Clarence Denton Tuska was born in New York City. At age eleven he was experimenting with wireless. In 1908 his family moved to Hartford and as a high school student had a complete amateur station in his house. Clarence used the call letters SNT.

In order to maintain his station and buy parts, Clarence made simple receiving equipment which he sold on consignment at a local hobby shop. That's where Hamilton Maxim, Percy's son got him to buy one of the receiving sets. A few days later Percy returned the receiving set to the hobby shop because it didn't meet with his approval. Needless to say, Clarence paid Maxim's household an apprehensive visit to find out what went wrong. Percy thought that Tuska's device wasn't powerful enough and put an order in for one which fit his requirement. The most concrete part of the visit, the bond of friendship that developed which had an overwhelming effect on amateur radio. Percy and Clarence were friends from that time on. In March, 1914 they came together to form the American Radio Relay League. [1]

MANUAL OF INSTRUCTIONS

GILBERT

ERECTOR

PART 2 MANUAL FOR SETS 7 and 8

PART 3 MANUAL PRIZE WINNING MODELS FOR ALL SETS

PRICE OF MANUALS No. 1 No. 2 No. 3 25c POSTPAID

PART 1
FOR SETS 1-6 INCLUSIVE

THE BOYS' CHOICE = " BE A GILBERT DIPLOMA BOY
GREAT FUN FOR BOYS = FUN AND FAME WHILE AT PLAY

THE A. C. GILBERT CO., NEW HAVEN, CONN., U. S. A.
IN CANADA
THE A. C. GILBERT-MENZIES CO., LIMITED, TORONTO, ONT.

Copyright 1919
The A. C. Gilbert Company, New Haven, Conn., U. S. A.

Sometime in late 1916 or early 1917 Percy acted as a judge in an A.C. Gilbert contest for boys. Gilbert was the maker of Erector sets and later American Flyer toy trains. During the judging, Percy preached the virtues of amateur radio until Gilbert asked Percy if he got into the amateur radio business would he be his consultant? Percy said no, but he knew someone that would be perfect for the job. He told Clarence:

> "Maxim told me of the conversation and suggested that I go to New Haven, see Mr. Gilbert and make any arrangement I wished, and threw in some fatherly advice. At the time (1916-1917), I was working my way through Trinity College in Hartford and any income-producing job looked good."

Have You Kept Pace With The Times?

Are you still operating the old style non-sensitive, poorly designed receiving apparatus? Or are you equipped with one of the new complete Gilbert Radio Outfits designed by an expert who was a Radio Officer in the U. S. Army during the war and who has incorporated all of the latest Army improvements in these new outfits.

Gilbert Radio Outfits are complete in every respect for transmitting, receiving or both. They are without doubt the latest and most scientific outfits. Each piece of apparatus is designed to operate in close harmony with every other part of the set.

The Loose Coupler used in the receiving sets is not the old Obsolete type that pulls out of the box but the new enclosed panel type. A highly sensitive instrument designed particularly to minimize distributed capacity and to eliminate dead-end losses. It is a very compact instrument in a quartered oak cabinet and is included in all Gilbert Radio Receiving Outfits.

These outfits can be purchased complete or each part separately. Our new Radio catalog No. 50 describes each one in detail. Write for it today.

THE A. C. GILBERT COMPANY

317 Blatchley Ave.,   New Haven, Conn.

IN CANADA
THE A. C. GILBERT-MENZIES CO., Limited
Toronto

ALWAYS MENTION QST WHEN WRITING TO ADVERTISERS

QST February 1920

Then came World War I. Clarence was working on a trench radio transmitter for Gilbert when he enlisted. There would be a job with Gilbert when he returned. Shortly after his return from the war Clarence became discouraged. He felt that Gilbert was using toy principles in the manufacture of amateur radio receivers. They parted company.

Clarence had an idea in the back of his mind to make inexpensive kits for young boys to learn electrical fundamentals.

"I talked this over with Maxim. We organized the C. D. Tuska Co., and he let me use a vacant room over his office (that is until I had a couple of helpers and we were just too noisy overhead. Remember he was the Silencer man!)." [2]

QST May 1920

Those electrical fundamental kits were a failure. However, some of the components in the kit had an application in radio. From this venture two products emerged, the Superdyne Receiver and the Tuska Tickler.

The "Superdyne" is a tuned radio amplifier in which a reversed feedback coil is used to prevent oscillation of the amplifier tube. While the Tuska Company expects to market complete Superdyne sets, we shall be glad to assist you if you are going to build your own. I should consider it a great favor if you would keep me personally advised as to what sort of success you have with it. [3]

He [Tuska] developed a differential capacitor called The Tuska Tickler. It had two sets of fixed plates and one interacting variable set. Regeneration was just coming into the radio art. This Tickler could be added to existing loose-couplers and audions producing capacitive feedback and regeneration. Armstrong [Edwin Howard] and his attorney convinced Tuska that the Tickler infringed on one of Armstrong's patents, so Tuska became the owner of an Armstrong license.

The C.D. Tuska Company started to expand in 1921. They moved from an attic room with three men and two boys, to a 1200-square foot loft. When orders for receivers escalated, the company moved again to a 3000-square foot factory type building. In January 1922, Tuska's employees numbered as follows: three boys, eight men and two stenographers. Shortly thereafter a sudden surge in radio sales hit. The company moved again into a larger building and employed eighty people. It wasn't too long after that when

RADIO WINS HIM MILLIONS

Clarence D. Tuska

Breesport, N. Y., June 13.—Two-not very long ago that Clarence D. Tuska wondered where his next meal would come from.

Today, at 30, he is a millionaire —one of the new radio millionaires.

Tuska first "got the radio bug" in 1912. He worked with models and experimented on the air late at night, using a small attic for his laboratory. One night Hiram Percy Maxim listened in on Tuska's broadcasting. Tuska thereupon became a protege of the great inventor.

In 1919, when he returned from the war, Tuska's workshop was moved from the attic. His many radio inventions brought him rapid success.

State Times Advocate Baton Rouge
06-12-24 p18

46

Tuska realized he had incorrectly estimated what he thought was a substantial growth in business. With his receiver production up and no new models planned, his company was in trouble. There was nothing written about the company decline until June 1923. An offering was made by a New York dealer that "five tons" of Tuska sets and parts were for sale. [4]

Clarence Tuska appeared not to have spent all of his time working. He must have had some leisure time which was captured in this newspaper article.

In mid-1925 Clarence called it quits in the radio business. It was getting to be an uncertain income.

Westinghouse sued, claiming Tuska had overstepped its license conditions by selling to jobbers and retailers.

Airplane Crashes Are New Cause for Suits

HARTFORD, Oct. 27 (A. P.)—Airplane collisions are a new cause of suits in the superior court, one having just been filed by Charles Davy of New Britain, who asks $1500 damages from Clarence D. Tuska of Hartford, as a result of a smash at Brainard Field, Sept. 13.

According to Davy's complaint, his airplane was slowly bumping along the ground, preparatory to letting a passenger out, when Tuska came speeding along in his plane on the ground, preparatory to a take-off, and ran head-on into Davy's machine, which was smashed so that it could not be repaired.

Boston Herald 10-28-25 p19

Once Clarence acquired the necessary releases from creditors (and twenty-four Stockholders) he sold his company. It was purchased by Atwater Kent in March, 1926.

With Westinghouse beginning to sue even makers of TRF radios for infringement of the Armstrong patent on grounds that their sets could regenerate by misadjustment of controls or by use of higher B-plus voltages; Kent probably felt it was worth having an Armstrong license in case Westinghouse won its point.

At the time of the sale, Clarence accepted a position at the Atwater Kent Company as assistant patent attorney. He held that position until 1935 when he moved to the RCA Corporation in New Jersey in a similar position. [5]

Clarence Tuska's name surfaces in original letters to James Millen. He had taken a toy company and turned it into a very successful electronics business. The letter was relating to the patent on the famous National PW dial drive assembly, often referred to as the HRO dial. This correspondence was dated April 17, 1939. [6]

The Radio Corporation of America promoted Clarence to Director of Patent Operations in 1947. [7] He continued to work at his job and in 1948 Clarence gave a speech at Princeton as part of a conference.

Princeton Rotary Club to Hear
Four Talks On Atomic Energy

PRINCETON—A series of four talks on the general subject, "Atomic Energy—What It Means to You," has been arranged for the Princeton Rotary Club. The first will take place next Tuesday at 12:45 in the Nassau Tavern, with the other three to follow at weekly intervals.

"The Legislative Aspect" will be discussed to open the series by Clarence D. Tuska. He is director of the Patent Department at RCA Laboratories.

Trenton Evening Times 02-04-48 p29

A change developed in Clarence Tuska's life sometime between late 1949 to early 1950. He met Edith, the woman who would become his wife. She was born in Scotland on December 2, 1895 and came to the U.S. in 1949. [8]

C. D. Tuska

Pioneer
In Radio
To Speak

Wireless was barely out of the cradle and so was Clarence D. Tuska when he started learning all about it.

It was about 1908, when he was 12, that Tuska made his first dot-dash transmitter from the spark coil of a Ford.

From there on in it was electronics—through the war, through the founding of his own manufacturing firm and right up until now. He's in Arlington to address the West Coast Division of the American Radio Relay League which is having its convention Saturday and Sunday at the Inn of the Six Flags.

This is a return trip to Texas for Tuska, who lives in Princeton, N.J. During World War I he established radio schools for the Signal Corps in Houston, at the University of Texas and at Love Field in Dallas.

Now retired, he likes to note that QST, the magazine for radio enthusiasts which he founded in 1915, then sold for a dime. "We couldn't even give it away at that price. Now it's a collectors' item and one recently sold for $12."

The organization, which he helped to establish, now numbers 225,000 internationally — a long way from the dot-dash days when you could only transmit up to 25 miles.

—Dallas News Staff Photo.

Clarence D. Tuska radioes a message.

Dallas Morning News 06-04-66 p.AP-12

48

Clarence retired from RCA in and around 1961. He did like to speak to groups. The next time newspapers give their attention to Clarence was in 1966.

The Tuska name appears in an obituary section of the a newspaper dated December 14, 1973.

Mrs. Edith W. Tuska

PRINCETON — Mrs. Edith W. Tuska, 78, of 401 Mercer Road died Thursday after a long illness.

Born in Scotland, she lived here since 1949 and was an army nurse in World War I.

Surviving are her husband, Clarence D. Tuska; a son, James W. of Hopewell Township; a daughter, Mrs. Joseph B. Jenks of Rockport, N.Y., two granddaughters, a brother and a sister.

Services will be held Saturday at 10 a.m. at the Kimble Funeral Home, 1 Hamilton Avenue, with the Rev. William L. Tucker officiating. Interment will be private. There will be no calling hours.

Trenton Evening Times 12-14-73 p44

Clarence Denton Tuska's long career came to an end. He died in July 1985 in Cranbury, Middlesex County, NJ 08512. He was 88 years old. [8]

Chapter 8: Epilog

After Percy's death, the American Radio Relay League, Federal Communications Commission and the 'ham' all tried to improve the world of amateur radio, each in their own way.

1938 -The Cairo Conference. Amateurs lose exclusive use of 40 meters. Must share with broadcasters. Federal Communications Commission presents hams with two new "UHF" bands, 2.5 meters (112 Mhz) and 1.25 meters (224 Mhz).

1940 -1941-War in Europe. International QSO's is severely limited. When United States enters the War, all amateur activity is suspended.

1942 -1945--Except for WERS (the War Emergency Radio Service) on 2 1/2 meters, no amateur operations take place.

1945 – Major battle over postwar frequency allocations. Parties involved: ARRL, David Sarnoff and Armstrong. The fight is over the low end of the VHF spectrum between 44-108 Mc. Of the three proposals made by the FCC, each party got something, except Sarnoff. FM is moved (over Armstrong's objections) from 42-50 to 88-108 Mhz. The FCC move amateur 2.5 meter band to 144-148 Mc (over the ARRL's objections). On November 15, 1945, amateurs were allowed back on the air, but only on 10 & 2 meters.

1945 - CQ magazine is first published.

1946 –Military leaves our HF bands as is over a year, except for 160 meters. The Military still use it for LORAN Radio navigation system. War surplus equipment was on its way to the ham community.

1947- Atlantic City Conference--Amateurs lose the top 300 khz of 10 meters (29.7--30), 14.35--14.4 Mhz on 20 meters. They however gain a new band at 15 meters (21.0--21.45 Mhz) in the future. FCC compensate hams for their loss allowing them 11 meter band (26.96--27.23 Mhz). The CK722 Transistor is invented by Bell Labs.

1948-Single Sideband is fully depicted in amateur publications. The FCC creates Class A & Class B CB radio between 460--470 Mc.

1951-Federal Communications Commission totally reorganizes the amateur license system. Class A, B, & C Licenses are replaced by the Advanced, General, & Conditional Class respectively. Three new license classes are created--Amateur Extra, Novice and Technician. The Technician Class is created for experimentation, not communication, and has privileges only above 220 Mc. Novices are given limited High Frequency CW sub bands, 75 watts and crystal control only. They may also use phone on 145--147 Mc. It is a 1 year, non renewable license.

1952- FCC allows phone operation on 40 meters, which had been CW only. The 15 meter band is opened. The Advanced Class is withdrawn from new applicants, although present holders can continue to renew, and the "exclusive" 75 & 20 meter phone bands are opened to Generals & Conditionals. Everyone, Conditional & above, has the same privileges.

1953- FCC starts issuing "K" calls in the 48 States due to increased ham population. The "W" calls were depleted.

1955-Technicians are given 6 meter privileges to help populate the band & encourage experimentation. ARRL and most hams opposed 2 meters for Technicians. Wayne Greene becomes editor of CQ magazine.

1956-1960--A transition comes into the ham shack. Transistors find their way first in power supplies, then audio sections, then receivers and finally QRP transmitters. Most ham equipment still use 100% tubes. SSB is catching on in popularity. In the 1960's, SSB pulls ahead of AM.

1957-Sputnik, is launched by the USSR. Amateurs copy its beacon on 20 & 40 Mhz.

1958 - Explorer is launched by the US. amateurs. Ham population reaches 160,000. Three time the number in 1946. FCC has to issue "WA" calls in the 2nd & 6th call areas, as the "W" & "K" 1x3 prefixes have run out. Slow Scan TV is first described in QST. In September, amateurs lose use of 11 meters. Class D CB is born.

1959 - Geneva Conference held, no major amateur changes. Technicians get the middle part of 2 meters (145-147 Mc), but not without some controversy over the purpose of the license. FCC restates their "experimental, not communication" policy.

1960 -Wayne Greene fired as CQ editor, forms 73 magazine.

1961- OSCAR I, the first amateur satellite, is launched.

1962 - CONELRAD is replaced by the Emergency Broadcast System. Amateurs no longer have to monitor 640 or 1240 khz while operating.

1963 - Amateur population is over 200,000, but CB licenses now outnumber hams.

1964 - Herbert Hoover dies at the age of 90. His strong support of amateur radio was invaluable. He lived long enough to see his son (Herbert Hoover, Jr, W6ZH) elected President of the ARRL.

1965 - OSCAR III & OSCAR IV allow 2 way QSO's via satellite.

1967-The FCC announced the new Incentive Licensing rules: over the next 2 years, General & Conditional operators would lose 50% of the 75-15 meter phone bands, the "First Class" idea was dropped, Advanced Class was reopened to new applicants, Extra & Advanced Class operators get exclusive sub bands on 80-15 and 6 meters, the Novice license term is doubled to two years, but Novices lose their 2 meter phone privileges, the FCC restates the "Technicians are experimenters, not communicators" policy, and states that the next license step for Novices is the General, not Technician, class.

1968 -FCC authorizes SSTV in the Advanced/Extra Class sub bands. Generals & Conditionals get SSTV later.

1969-The FCC removes the ability for a Technician to hold a Novice license at the same time. The ARRL announces a new policy, they now consider Technicians to be communicators and petition FCC to give them full VHF privileges, 10 meter segment from 29.5-29.7 Mhz, and Novice CW sub bands.

1970-The amateur population is 250,000. Two meter FM is starting to boom. New terms emerge, Mhz for Mc and Khz for Kc.

1971-Japanese are starting to dominate amateur markets. National, Hammarlund, Hallicrafters and Gonset beginning to dwindle. Drake, Ten-Tec, Heathkit and Collins were still doing fine.

1972 – Two meter FM band plan was announced, 146.52 was chosen as the national simplex frequency. FCC released the first repeater rules and expanded the Technician 2 meter allocation to 145-148 Mhz. They also relaxed mobile logging requirements.

1975-1976 - New repeater sub band is established at 144.5-145.5 Mhz. Technicians now have 144.5-148 Mhz on 2 meters and have Novice privileges. Novices are given a power increase to 250 watts. The "mail order" Technician license is eliminated. The Conditional class is abolished.

1977 - FCC expands CB radio from 23 to 40 channels. Hams purchase "obsolete" 23 channel CB sets converting them to 10 meters.

1978-Technicians finally get all privileges above 50 Mhz, and can obtain a RACES Station authorization. Novice license is made renewable. Amateur population is at 350,000. "Packet" radio first appears on an experimental basis.

1979 -World Administrative Radio Conference, or WARC-79, takes place in Geneva. ARRL, IARU & other groups have been preparing for years. American hams lose nothing and gain three new bands at 10, 18, & 24 Mhz, which are phased in during the next 10 years.

1980-FCC authorizes ASCII on ham bands. Packet is starting to grow.

1982-The "Goldwater" Bill is passed. It allows the FCC to set industry standards regarding RFI.

1983 - Owen Garriott, W5LFL, becomes the first amateur to operate from the Space Shuttle. Another "Code Free" license idea pops up. Proposal dropped after huge opposition.

1984-A ten year license replaces the 5 year. FCC stopped giving examinations, turning the duty over to the new Volunteer Examiner Program. The HF phone bands are expanded. The amateur population is up to 410,000.

1985-State and local rules which restrict amateur antennas must now comply with the FCC's new policy, expressed in PRB-1.

1987-Novices & Technicians get 10 meter SSB privileges from 28.3-28.5 Mhz. Novices also get phone operation on portions of 220 & 1296 Mhz. Element 3 written exam is broken into 2 segments--3A (Technician) and 3B (General). Technicians who passed their exam prior to March 1987 get permanent credit towards the General written exam.

1989-Amid growing calls for a code free license, the ARRL comes out in favor of one.

1990-1991--MARS operations increased as amateurs became involved in Operation Desert Shield/Storm.

1991-Amateur Radio gets its first code free license--the "No Code Technician". "Regular" Technicians are renamed "Technician Plus."

1991-1998--Amateur Radio grows from 500,000 to over 710,000 hams. ARRL is at its highest membership ever. Schoolchildren talk with hams in space. Our Public Service activities are wanted & appreciated. And Amateur Radio looks forward to the next Millennium, confident that it will evolve and grow.

2004 – FCC and BPL (Broadband over Power Line) rules.

2007 – Elimination of the Morse code requirement for amateur radio operator licenses.

2011—New rule changes to the vanity call sign system and call signs for Amateur Radio clubs.

2012-- Amateurs have number of new privileges on the 60 meter band. A boost in effective radiated power from 50 to 100 W, as well as the ability to use CW and certain digital modes.

2013 -- The FCC has authorized the Maxim Memorial Station W1AW to also use special call sign W100AW during 2014, the ARRL's centennial year.

Acknowledgements

AI7H, Ed Stuckey, Idaho Section Manager, who helped me get this project off the ground.

N7JU, John Hollar Jr., for helping with the editing.

Mary Smith who helped with the final edit and gave me that gentle nudge to keep writing.

ARRL who has a wonder mountain of historic material called QST Magazine.

Source Notes

Chapter 1: Those who came before

1. http://en.wikipedia.org/wiki/Alessandro_Volta
2. http://inventors.about.com/od/tstartinventions/a/telegraph.htm
3. Before Spark, Gil McElroy VE1PKD, QST, January 1994, p.57.
4. http://en.wikipedia.org/wiki/Hans_Christian_%C3%98rsted
5. http://inventors.about.com/od/tstartinventions/a/telegraph.htm
6. http://en.wikipedia.org/wiki/Andre-Marie_Ampere
7. http://inventors.about.com/library/inventors/blohm.htm
8. http://inventors.about.com/od/hstartinventors/a/Joseph_Henry.htm
9. http://en.wikipedia.org/wiki/Harrison_Gray_Dyar
10. http://inventors.about.com/od/tstartinventions/a/telegraph.htm
11. http://en.wikipedia.org/wiki/Samuel_Morse
12. The Man before Marconi, Joseph R. Rebo, W2OEU, QST, August 1948, p. 42.
13. Before Spark, Gil McElroy VE1PKD, QST, January 1994, p.57.
14. http://en.wikipedia.org/wiki/James_Bowman_Lindsay
15. http://corporate.westernunion.com/History.html
16. http://en.wikipedia.org/wiki/Guglielmo_Marconi
17. The Man before Marconi, Joseph R. Rebo, W2OEU, QST, August 1948, p. 43.
18. http://en.wikipedia.org/wiki/Samuel_Morse
19. http://en.wikipedia.org/wiki/Alexander_Graham_Bell
20. https://www.google.com/#q=David+E.+Hughes+
21. http://history1900s.about.com/od/people/a/Tesla.htm
22. http://en.wikipedia.org/wiki/Heinrich_Hertz
23. http://en.wikipedia.org/wiki/Invention_of_radio

Chapter 2: Spark-gap, and the Hams

1. Who Invented Radio, Larry Kahaner, 73 Magazine, December 1980, p.40.
2. Desoto, Clifton B., 200 Meters and Down, Rumford Press, Concord N.H., p.18
3. http://en.wikipedia.org/wiki/Lee_de_Forest
4. Desoto, Clifton B., 200 Meters and Down, Rumford Press, Concord N.H., p.20.
5. http://en.wikipedia.org/wiki/Alexanderson_alternator
5. http://www.ieeeghn.org/wiki/index.php/Milestones:Poulsen-Arc_Radio_Transmitter,_1902.
6. Desoto, Clifton B., 200 Meters and Down, Rumford Press, Concord N.H., p.20.
7. Remembering Hugo Gernback, Gil McElroy, VE1PKD, *QST Magazine*, February 1995, p.37.

8. Fifty Years of ARRL, Newington, Conn. ARRL 1965, p.8.
9. Desoto, Clifton B., 200 Meters and Down, Rumford Press, Concord N.H., p.22-30.
10. Jim Maxwell W6CF, "Amateur Radio: 100 Years of Discovery," *QST Magazine*, January 2000, n.p.
11. Remembering Hugo Gernback, Gil McElroy, VE1PKD, QST Magazine, February 1995, p.38.
12. Jim Maxwell W6CF, "Amateur Radio: 100 Years of Discovery," *QST Magazine*, January 2000, n.p.
13. The Father of Modern Radio, Morton Eisenberg, K3DG, QST Magazine, May 1991, p. 49.
14. Desoto, Clifton B., 200 Meters and Down, Rumford Press, Concord N.H., p.37.
15. Jim Maxwell W6CF, "Amateur Radio: 100 Years of Discovery," *QST Magazine*, January 2000, n.p.

Chapter 3: Hiram Percy Maxim – The Early Years

Schumacher, Alice Clink, Hiram Percy Maxim, Electric Radio Press, Cortez, Colorado, 1998, p.2-17.
Maxim, Hiram Percy, A Genius in the Family, Benediction Classics, Oxford, England, 2010, p. 139-144.
Schumacher, Alice Clink, Hiram Percy Maxim, Electric Radio Press, Cortez, Colorado, 1998, p.2-17.
Maxim, Hiram Percy, Horseless Carriage Days, Dover Productions, Inc., New York, 1962. p. 57-59.
Schumacher, Alice Clink, Hiram Percy Maxim, Electric Radio Press, Cortez, Colorado, 1998, p.24.
Maxim, Hiram Percy, Horseless Carriage Days, Dover Productions, Inc., New York, 1962. p. 57-59.
Schumacher, Alice Clink, Hiram Percy Maxim, Electric Radio Press, Cortez, Colorado, 1998, p.23-44.
Maxim, Hiram Percy, Horseless Carriage Days, Dover Productions, Inc., New York, 1962. p. 139-144.
Schumacher, Alice Clink, Hiram Percy Maxim, Electric Radio Press, Cortez, Colorado, 1998, p.48-52

Chapter 4: Hiram Percy Maxim and the ARRL

1. Schumacher, Alice Clink, Hiram Percy Maxim, Electric Radio Press, Cortez, Colorado, 1998, p.52-53.
2. Desoto, Clifton B., 200 Meters and Down, Rumford Press, Concord N.H., p.37-41.
3. The Father of Modern Radio, Morton Eisenberg, K3DG, QST Magazine, May 1991, p. 49-50.
4. Schumacher, Alice Clink, Hiram Percy Maxim, Electric Radio Press,

Cortez, Colorado, 1998, p.54-56.
5. Clifton B., 200 Meters and Down, Rumford Press, Concord N.H., p.46.
6. Remembering Hugo Gernback, Gil McElroy, VE1PKD, *QST Magazine*, February 1995, p.39.
7. Schumacher, Alice Clink, Hiram Percy Maxim, Electric Radio Press, Cortez, Colorado, 1998, p.56-57.
8. Clifton B., 200 Meters and Down, Rumford Press, Concord N.H., p.50.
9. Schumacher, Alice Clink, Hiram Percy Maxim, Electric Radio Press, Cortez, Colorado, 1998, p.58-59.

Chapter 5: War Over – Back on the Air

1. DeSoto, Clifton B., 200 Meters and Down, Rumford Press, Concord N.H., p.53.
2. The Father of Modern Radio, Morton Eisenberg, K3DG, QST Magazine, May 1991, p. 50.
3. Schumacher, Alice Clink, Hiram Percy Maxim, Electric Radio Press, Cortez, Colorado, 1998, p.60-61.
4. The Father of Modern Radio, Morton Eisenberg, K3DG, *QST Magazine*, May 1991, p. 50-51.
5. Schumacher, Alice Clink, Hiram Percy Maxim, Electric Radio Press, Cortez, Colorado, 1998, p.61.
6. DeSoto, Clifton B., 200 Meters and Down, Rumford Press, Concord N.H., p.57.
7. Schumacher, Alice Clink, Hiram Percy Maxim, Electric Radio Press, Cortez, Colorado, 1998, p.62-63.
8. Jim Maxwell W6CF, "Amateur Radio: 100 Years of Discovery," *QST Magazine*, January 2000, n.p.
9. Amateur Radio Licensing: A Seven Decade Overview, Neil D. Friedman, N3DF, QST Magazine, March 1985, p. 47.
10. http://en.wikipedia.org/wiki/QSL_card
11. Schumacher, Alice Clink, Hiram Percy Maxim, Electric Radio Press, Cortez, Colorado, 1998, p.64-69.
12. DeSoto, Clifton B., 200 Meters and Down, Rumford Press, Concord N.H., p.66-77.
13. The Final Days of Ham Spark, Harry R. Hyder, W7IV, *QST Magazine*, March 1992, p. 32.
14. Schumacher, Alice Clink, Hiram Percy Maxim, Electric Radio Press, Cortez, Colorado, 1998, p. 79-84.
15. DeSoto, Clifton B., 200 Meters and Down, Rumford Press, Concord N.H., p.93.
16. Schumacher, Alice Clink, Hiram Percy Maxim, Electric Radio Press, Cortez, Colorado, 1998, p. 84-88.

Chapter 6: A New Beginning

1. http://scholarship.law.marquette.edu/cgi/viewcontent.
cgi?article=4396&context=mulr
2. Schumacher, Alice Clink, Hiram Percy Maxim, Electric Radio Press, Cortez, Colorado, 1998, p. 90-92.
3. DeSoto, Clifton B., 200 Meters and Down, Rumford Press, Concord N.H., p.120-122.
4. http://www.rollanet.org/~n0klu/Ham_Radio/History%20of%20Ham%20 Radio.pdf
5. DeSoto, Clifton B., 200 Meters and Down, Rumford Press, Concord N.H., p.124-125.
6. http://www.rollanet.org/~n0klu/Ham_Radio/History%20of%20Ham%20 Radio.pdf
7. Schumacher, Alice Clink, Hiram Percy Maxim, Electric Radio Press, Cortez, Colorado, 1998, p. 94-95.
8. DeSoto, Clifton B., 200 Meters and Down, Rumford Press, Concord N.H., 1936, p.142-143.
9. Schumacher, Alice Clink, Hiram Percy Maxim, Electric Radio Press, Cortez, Colorado, 1998, p. 102-109.
10. http://www.hamradiomarket.com/articles/SSBHistory.htm
11. http://en.wikipedia.org/wiki/Edwin_Howard_Armstrong
12. The Father of Modern Radio, Morton Eisenberg, K3DG, *QST Magazine*, May 1991, p. 51.
13. Schumacher, Alice Clink, Hiram Percy Maxim, Electric Radio Press, Cortez, Colorado, 1998, p. 114-118.
14. Tuska, Clarence D., Inventors and Inventions, McGraw-Hill Book Company, New York, NY., 1957, p. 116-117.

Chapter 7: Clarence D. Tuska

1. Clarence D, Tuska: 1896-1985, W1RW, *QST Magazine*, September 1985, p.45.
2. Alan Douglas, "Manufactures of the 1920s, vol. 3." page 201-203.
3. The Superdyne Receiver, C. D. Tuska, *QST Magazine*, November 1923, p. 7-12.
4. Terry, W5OAS, http://www.radioera.com
5. Alan Douglas, "Manufactures of the 1920s, vol. 3." page 201-203.
6. http://www.isquare.com/millen/bio_rem/merrill.html
7. Alan Douglas, "Manufactures of the 1920s, vol. 3." page 203.
8. http://www.newspaperarchive.com

Bibliography

Books

DeSoto, Clinton B. (1936) *Two Hundred Meters and Down*, Concord, New Hampshire, The Rumford Press.

Maxim, Hiram Percy, (2010) *A Genius in the Family*, Oxford, England, Benediction Classics. ISBN: 978-1-84902-619-2.

Tuska, Clarence D., (1957) *Inventors and Inventions*, New York, New York. McGraw-Hill Book Company, Inc.

Schumacher, Alice Clink, (1998) *Hiram Percy Maxim*, Cortez, Colorado, Electric Radio Press, Inc., ISBN: 0-9663433-C-1

Magazines

Friedman, Neil D., N3DF, Amateur Radio Licensing: A Seven Decade Overview, *QST Magazine*, March 1985.

Hyder, Harry R., W7IV, The Final Days of Ham Spark, *QST Magazine*, March 1992.

Kahaner, Larry, Who Invented Radio, *73 Magazine*, December 1980.

Maxwell, Jim W6CF, "Amateur Radio: 100 Years of Discovery," *QST Magazine,* January 2000.

McElroy, Gil, VE1PKD, Before Spark, *QST Magazine*, January 1994.

McElroy, Gil, Remembering Hugo Gernback, VE1PKD, *QST Magazine*, February 1995.

Eisenberg, Morton, K3DG, The Father of Modern Radio, *QST Magazine*, May 1991.

Rebo, Joseph R., The Man before Marconi, W2OEU, *QST Magazine*, August 1948.

Author

Photo by John N7JU

Larry Telles, K6SPP passed his Tech exam on April 15, 1958. He had gone to work for Pacific Telephone Company a year earlier. He spent the last 16 years of his 30 years with Bell as a technical instructor in voice and data. After taking early retirement in 1987 Larry went to work for Network Equipment Technologies in Redwood City, California, Stratacom, and ITT Tech before moving to the Coeur d'Alene area in 1999. While at ITT Tech he taught operating systems, data networking, and multimedia.

Larry is the past president of the Idaho Writers League, Coeur d'Alene Chapter, and webmaster of four websites, one of which K7ID.ORG. When he is not talking on the radio, taking pictures, or drawing, Larry is busy in his other creative mode, writing. In 2014 he published his second book and produced a DVD. Once or twice each semester he teaches several short evening continuing education classes at North Idaho College, in Post Falls.

Larry promotes the ARRL as an ARRL Instructor and Volunteer Examiner as the team leader. He will use this material in his presentation at the Boise Hamfest in April, 2014.

www.ingramcontent.com/pod-product-compliance
Lightning Source LLC
Chambersburg PA
CBHW071850020426
42331CB00007B/1935